Primer of
Cancer
Management

Primer of Cancer Management

Jacob J. Lokich, M.D.

New England Deaconess Hospital
Sidney Farber Cancer Institute
Harvard Medical School
Boston, Massachusetts

G. K. Hall & Co.
Boston

G. K. Hall & Co.
Medical Publications Division
70 Lincoln Street
Boston, Massachusetts 02111

78 79 80 81/8 7 6 5 4 3 2 1

Lokich, Jacob J
  Primer of cancer management.

  Includes bibliographical references and index.
  1. Cancer.   2. Therapeutics.   I. Title.
  [DNLM: 1. Neoplasms—Therapy.   QZ266 L836p]
  RC270.8.L64   616.9'94'06   78-16752
  ISBN 978-94-010-9680-5   ISBN 978-94-010-9678-2 (eBook)
  DOI 10.1007/978-94-010-9678-2

Contributors

George L. Blackburn, M.D., Ph.D.
Associate Professor of Surgery
Director of Nutrition Support Service
New England Deaconess Hospital
Boston, Massachusetts 02115

James Bruckman, M.D.
Department of Radiotherapy
Harvard Medical School
Boston, Massachusetts 02115;
Clinical Fellow, Joint Center for Radiation Therapy
Harvard Medical School
Boston, Massachusetts 02115

Christopher C. Gates, M.D.
Associate in Medicine (Psychiatry)
Peter Bent Brigham Hospital
Boston, Massachusetts 02215;
Instructor in Psychiatry
Harvard Medical School
Boston, Massachusetts 02115

Jacob J. Lokich, M.D.
Oncologist
Sidney Farber Cancer Institute;
Assistant Professor of Medicine
Harvard Medical School,
Chief, Section of Medical Oncology
Department of Medicine
New England Deaconess Hospital
Boston, Massachusetts 02115

Althea Pisinski, R.N.
Former Tumor Immunology Nurse
Sidney Farber Cancer Institute
Boston, Massachusetts 02115

Geoffrey K. Sherwood, M.D.
Instructor in Medicine
Harvard Medical School,
Boston, Massachusetts 02115;
Division of Hematology-Oncology
Beth Israel Hospital
Boston, Massachusetts 02215

David Valerio, M.B.Ch.B.
Research Fellow in Surgery
Harvard Medical School
Boston, Massachusetts 02115

Ralph Weichselbaum, M.D.
Assistant Professor of Radiotherapy
Harvard Medical School
Boston, Massachusetts 02115;
Attending Radiotherapist
Joint Center for Radiation Therapy
Harvard Medical School
Boston, Massachusetts 02115

Acknowledgements

The idea for this primer was inspired by the patients who have been a part of my life. My mentors, Tom Frei and Bill Moloney, who educated me in the science and the art of cancer medicine, have encouraged me to fulfill this idea. Several people have shared cancer care with me, the dedicated nurses, the clinical oncologists, the students, and the house staff at the New England Deaconess Hospital and the Sidney Farber Cancer Institute. For all of these associations I am grateful personally and professionally.

Dedication

For patients and families who shared their lives and their tragedies with me and in all ways made my life more meaningful.

Preface

Cancer is a general term which applies to a broad spectrum of malignant diseases. For most patients, the diagnosis of cancer evokes images of intractable pain, mutilation, and impending death. Many of these feelings are derived from distortions of one's experiences with relatives or friends, or are based on misconceptions or myths about the disease itself. Fortunately, technologic and therapeutic advances have created a more optimistic future for the cancer patient. Health care teams, specializing in oncology, are defining and attending to the special needs of cancer patients. They are developing cancer management profiles for each patient which involve not only the primary diagnostic and therapeutic aspects of the disease but also its secondary effects on physiologic and psychologic processes.

The purpose of this primer is to function as a general guide for managing all stages of cancer—from the early curable tumor to advanced metastasis. Therapy for such patients may require surgery, chemotherapy, radiation therapy, or a combination of modalities including immune therapy. This text will focus on cancer care and the role of the staff in the following general areas: (1) the communication of the diagnosis and the characteristics of the disease; (2) the application of the therapeutic modalities; (3) the management of secondary symptoms such as pain; and (4) the development of support services such as psychological counseling.

The concept of total cancer care is emphasized throughout this text and will be pertinent to the professionals who affect the lives of cancer patients—physicians, nurses, social workers, dieticians, pharmacists, clergymen, psychologists, rehabilitation specialists, and technologists. An essential therapeutic ingredient in cancer care is the compassionate understanding of the patient's anguishes and fears. The health care team can work together to function as a cohesive, interdependent group of educators, therapists, and in many ways, friends.

Table of Contents

Primer of
Cancer
Management

Section I

Cancer Therapy: Background and Patient Interaction

1

Concepts in Cancer Therapeutics

Chapter 1
Concepts in
Cancer Therapeutics

1.0

Introduction
Cancer therapy spans the disciplines of surgery, radiation therapy, chemotherapy, and immunotherapy. To these tumor-specific therapies may be added the general modalities of support therapy, including rehabilitation, and psychological therapy, and the multiplicity of non-specific therapies, such as pain management and transfusion. In more general terms, cancer therapy may be subdivided into two broad categories: curative therapy and palliative therapy. The former therapy usually consists of surgically excising the malignant lesion with the expectation of achieving cure. Radiotherapy has become recognized as another curative measure, particularly in treating Hodgkin's disease and carcinoma of the cervix. Chemotherapy may have a role in curative therapy but is more properly applied in adjuvant therapy in connection with either radiation or surgery.

Palliation therapy relieves symptoms or, in special circumstances, prevents their development. Generally, palliation therapy is provided by local radiation therapy to areas of symptomatic disease or is included in the non-specific support therapy for the secondary effects of the tumor, e.g. transfusion for anemia. Chemotherapy is usually placed in a category of a palliative modality as it has always been used in cases of advanced disease when all other therapies have failed. This chapter will define some of the newer concepts of therapy and will explore the choice of therapeutic modality, the criteria for timing of therapy, the determinants of response to therapy, and finally, the evaluation of the effects of therapy.

2.0

Therapeutic Roles of the Three Modalities
Surgery is involved in establishing the diagnosis by a biopsy procedure, and is the modality by which total tumor extirpation promotes the possibility of cure. It is also used as a palliative measure to relieve pain, bypass obstruction, and "debulk" the host-tumor burden in localized or regional dis-

ease. Diseases that do not require primary surgery are the hematologic malignancies, such as leukemia and lymphoma.

Radiation therapy, like surgery, is applied in local or regional disease and is used in metastatic disease to alleviate localized symptoms secondary to metastases. At times, radiation therapy may be used as a potential curative therapy for advanced stage disease, particularly if surgical procedures would provoke major morbidity. For example, extensive carcinoma of the breast, with a locally inoperable lesion, requires radiation treatment. The major role for radiation therapy, however, is in the palliation of localized symptomatology, such as bone pain or venocaval obstruction.

Chemotherapy or "systemic" therapy uses cytotoxic drugs as well as immune stimulant agents to promote tumor cell kill. Systemic therapy is defined by the capability of distribution throughout the body and thus differs from the local or regional therapies of surgery and radiation. In contrast to surgery and radiation therapy, chemotherapy is generally not employed for the treatment of local or regional symptomatic disease unless the specific tumor type is particularly responsive to a chemotherapeutic regimen. An exception is infusion and perfusion chemotherapy which will be discussed in Chapter 14.

One may define and categorize tumors according to their responsiveness to chemotherapy (Table 1.1) and may define the role of drug therapy in patient management in general terms. These categories apply only to advanced metastatic can-

Table 1.1
Tumor Responsiveness to Chemotherapy and Radiotherapy

Response Level	Chemotherapy	Radiotherapy
A. Exquisitely Responsive	Ovary	Lymphoma
	Breast	Head & Neck
	Genito-Urinary Cancer (testicular, prostate)	Cervix & Uterus
B. Moderately Responsive	Sarcoma	Lung Cancer
	Head & Neck	Rectum
	Oat Cell (lung)	Breast
	Osteogenic Sarcoma	Prostate
C. Unresponsive	Renal Cancer	Melanoma
	Lung Cancer	Renal Cell
	Melanoma	Pancreas
	Colon	Sarcoma
	Pancreas	

cer. Exquisitely responsive tumors include breast cancer, ovarian cancer, genito-urinary malignancy, and the lymphomas and leukemias. The moderately responsive tumors include sarcoma, some forms of lung cancer, gastric cancer, and bone tumors. In the category of generally unresponsive tumors are malignant melanoma, most lung cancers, and renal cell carcinoma. Pancreatic carcinoma and colon cancer can also be contained in this category as no more than one in five patients responds (and then only partially so) to these treatments. Such tumor responsiveness to drug therapy does not necessarily mean curability or even prolonged survival, but does reflect objective decrease in tumor size as a consequence of treatment.

3.0

Choosing the Therapeutic Modality

How does one choose the specific modality of therapy? Actually, multiple modalities of therapy are often employed concomitantly and the interdigitation of surgery, radiation therapy, and chemotherapy forms the basis for the adjuvant therapeutic approach to early cancer. This aspect of cancer care will be discussed more fully in the section on adjuvant therapy. Before we discuss integrated forms of cancer care, it would be helpful to identify some basic problems in establishing modes of treatment.

One of the basic rules of therapy in metastatic disease is to define and separate local complications from systemic or constitutional effects of the tumor. The former are approached either surgically or with radiation therapy and the latter by systemic chemotherapy.

A second issue involved in choosing the therapy is establishing the therapeutic index of the treatment modality. One must balance the therapeutic advantage or potential for tumor regression against the adverse effects of the treatment. A high therapeutic index implies a high likelihood of response at minimal cost. The decision to adopt a specific therapy is simple when its side effects are minimal and it promotes relief of symptoms or prolongation of life. However, when treating patients with metastatic disease with whatever modality, the therapeutic index is often marginal or unpredictable.

The use of potentially toxic therapy requires an individual determination of the indication for and the objectives of such therapy. The specific instances in which therapy should be employed are detailed in Table 1.2. For those patients who have exquisitely responsive tumors for which an effective therapy is known, the indication for treatment is clear. In the absence of these conditions, however, the application of therapy must be conditioned by either the presence of measurable disease, the presence of symptomatic disease, or the presence of clinically progressive disease. Without one of these three indications, it is hazardous and unwarranted to employ chemotherapy. The presence of measurable disease promotes the opportunity to determine the effectiveness of therapy early on and to discontinue therapy in the absence of an effect on the tumor, or conversely, to continue and perhaps intensify therapy when an anti-tumor effect is observed. Less objective but symptomatic disease, particularly constitutional symptoms such as fever, weight loss, and generalized nausea, is an appropriate reason for the introduction of therapy. Anorexia or a major change in functional status are also adequate reasons.

Table 1.2
Rules for When to Treat Advanced Incurable Malignant Disease

---

(1) Known responsiveness to therapy

(2) Measurable disease with or without known responsiveness to treatment

(3) Symptomatic disease (regional or constitutional) limiting life style or performance functioning

(4) Progressive disease or deterioration with limited and possibly measurable life expectancy, i.e. predictable future and acceptable risk

---

Many oncologists withhold all forms of therapy from the patient with asymptomatic metastases even when this disease is measurable and progressive. This "no-hands" approach admits hopelessness—an attitude that may be considered standard therapy but which is debatable. The major argument against this nihilistic posture is that the transition from asymptomatic to symptomatic disease invariably implies a larger host-tumor burden and therefore a decreased likelihood of response to treatment. Finally, the presence of rapidly progressive or evolving disease, even in the absence of symptoms or of specifically measurable parameters, may justify the introduction of treatment. Patients who do not fulfill one of these criteria may still receive tumor-specific treatment solely because their life expectancy is predicted to be short.

4.0
Adjuvant or Multiple Modality Therapy
Therapy introduced in addition to surgery is adjunctive or supportive treatment. For example, the goal in adding radiation therapy or chemotherapy to primary surgical therapy is to promote regional or distant disease control, respectively, in order to prevent a relapse of the disease. This multiple modality approach allows many forms of therapy to play important and primary roles in early disease. The proven efficacy of radiation therapy and chemotherapy in promoting cure in tumors such as uterine, ovarian, and breast cancer has established an important if not crucial role for these previously considered ancillary modalities.

Let us consider for a moment the conceptual rationale for adjuvant therapy. Historically, adjuvant forms of treatment were introduced because of the known hematogenous disbursement of tumor cells during surgical manipulations of tumors. It was demonstrated repetitively that at the time of mastectomy or colectomy, cellular debris and clonogenic* tumor

---

*Clonogenic describes the capability of forming a clone or colony upon implantation within a host organ such as the liver or the lung.

cells could be identified in the circulating blood volume. Thus, many of the initial forms of adjuvant therapy involved the actual intra-operative injection of the drug as well as postoperative treatment for a short period of time.

More recently, it has become clear that tumor manipulation does not contribute to the pathophysiology and development of distant metastasis. It is more likely that micrometastases are dispersed occultly during primary tumor growth and are simply subclinical at the time of primary tumor discovery. Such metastases subsequently grow and become clinically recognizable at a stage of advanced disease some time after the initial diagnosis and surgical treatment.

These micrometastases are more responsive to effective forms of chemotherapy and radiation therapy than macrometastases. The microscopic lesions contain a large percentage of cells in the growth cycle which are responsive to toxic interruption of DNA synthesis in contrast to large tumors in which the bulk of the tumor cells are in the resting phase. The effectiveness of multimodality therapy has not been proven completely, but some leads are developing. A review of the status of adjuvant therapy in a variety of tumors is indicated in Table 1.3. In these studies, the objective is to decrease the incidence of recurrent disease. The decreased recurrence of disease, however, may be simply a delay in recurrence and the overall cur-

Table 1.3
Present Status of Adjuvant Therapy In Selected Tumors

Tumor	Therapy	Impact on Natural History
Breast	Cytoxan, Methotrexate, Fluorouracil (CMF) L-Phenylalanine mustard (LPAM)	Improved survival and decreased relapse rate in premenopausal patients
Ovary	LPAM or Colloidal Gold	Unproven but suggested trend
Colon	5-Fluorouracil (FU)	No effect
Lung	Chemotherapy	No effect
	Radiation	No effect
Rectum	Radiation	Decreased recurrence
Melanoma*	Immune Therapy	Suggested decreased recurrence rate*

*Experimental design disputed

ability may not be affected. A distinction may be made between prolonging the time until relapse, and prolonging survival or achieving curability. It has been difficult to establish the survival impact of adjuvant treatment because the clinical trials involve patients with early disease and a long period of follow-up is necessary.

Prognostic Determinants

The most critical determinant of prognosis in the cancer patient is the stage of the disease. In the staging process the extent of the tumor is defined in anatomical terms which is related in turn to the quantitative host-tumor burden. Other prognostic factors are clearly of major importance: 1) the biological characteristics of the tumor as reflected in its growth rate and invasive properties; 2) the basic health of the host including age, vital organ function (renal, cardiac, hepatic), and immunologic capability; and 3) the sensitivity or responsiveness of the tumor to therapeutic modalities.

Before we consider these factors, it is important to further analyze the process of staging. Staging is useful because it is a prognostic determinant and functions to focus and aid therapeutic decisions. Early stage cancer with an excellent prognosis is by and large managed with the surgical approach alone. More advanced disease may be managed solely with radiation therapy or chemotherapy. Examples of some of the various staging systems as they apply to a variety of tumors are illustrated in Table 1.4. The T (for tumor), N (for node), and M (for metastasis) is the staging system employed for a variety of tumors including breast cancer, lung cancer, and some gastrointestinal tumors. It is an effective means of staging as well as of describing the anatomic distribution of the tumor. T is often separated into two or three categories, depending upon the primary tumor size. N is also separated into various categories depending upon the size and fixation of the nodes. For example, a T2 N1 M0 breast cancer represents a lesion that is more than 2 and less than 5 centimeters in size within the breast structure with palpable nodes in the axilla measuring less than 2.5 cm and presenting no evidence of distant metastases.

The other staging systems listed in Table 1.4 are basically

5.0

Table 1.4
Staging Systems in Cancer

System	Tumor
1. TNM (Tumor, Node, Metastases)	Lung, Breast
2. I, II, III	Ovary, Cervix, Hodgkin's
3. A, B, C	Prostate, Bladder
4. Duke's Classification (and modification)	Colon, Rectum

three-component systems which apply to local (Stage I or A), regional (Stage II or B), and disseminated (Stage III or C) disease. The major purpose of such staging is to establish on an anatomical basis, the prognosis which should be independent of other features of the disease.

Other prognostic determinants are often tumor-specific in orientation. For example, the circulating level of human chorionic gonadotrophin is an important prognostic determinant of testicular and choriocarcinoma, independent of the stage of the disease. Another important tumor-related prognostic feature is the pathologic characteristics of the tumor. Pathologists often use Broder's classification system to grade the tumor. Well-differentiated or grade I tumors have an excellent prognosis which is independent of stage. The grade IV or less well differentiated so-called anaplastic tumors have a much poorer prognosis generally also independent of the stage of the disease. Not all tumors are graded but head, neck, bladder, and ovarian cancer are some of the tumors which are most commonly incorporated into a pathologic grading classification.

6.0 **Determinants of Response to Therapy**
The clinical features which determine the potential for response to either drug or radiation therapy are important components in the formulation of a therapeutic plan (Table 1.5). Two general categories of response determinants may be established: host factors and tumor factors. There are, in addition, specific features of the drug or radiation therapy which contribute to the conditions for response and these factors will be discussed in the respective chapters covering those therapeutic modalities. The host factor of major importance is performance status which is a quantitative translation of the presence of symptoms. The patient with symptoms and with a life expectancy of less than 8 weeks, for example, is statistically less likely to respond to therapy in spite of the fact that the tumor is exquisitely responsive to therapy. Additional host

---

Table 1.5
Determinants of Response to Therapy

Host Factors	Performance Status
	Prior Therapy
	Age
	Marrow Reserve
Tumor Factors	Site of Metastases
	Quantitative Tumor Burden
	Sensitivity to Therapy
	Growth Rate or Growth Fraction

features may affect the response by restricting the amount of drug which may be delivered to the host, such as prior therapy, age, and marrow reserve. All of these features have a negative impact on the likelihood of response to therapy; prior therapy particularly does because of a cumulative adverse effect on the host. The general statement may be made that if a tumor responds initially and subsequently recurs or reappears in the face of therapy, the likelihood of a second response to the same or new therapy is small. The corollary is that once a tumor recurs after primary radical or definitive therapy, the likelihood that a cure may be obtained is small.

A multiplicity of tumor characteristics affect potential responsiveness to therapy. One of the prominent features of tumors is the site of distribution of metastases. In breast cancer, a tumor may metastasize to the lung, the liver, bones and skin, as well as to the brain and other unusual sites. The potential for response is different within each organ site. Breast cancer metastatic to bone is more responsive than metastases in the liver. The reason for such differential responsiveness is not at all clear. Another tumor characteristic relating to responsiveness to therapy is found in tumors metastatic to "pharmacologic sanctuaries": These sanctuaries are areas of the body to which drug distribution may be restricted because of decreased blood flow or natural barriers. For example, tumors metastatic to the brain are not generally affected by drugs because, with few exceptions, drugs normally distributed systemically throughout the body do not penetrate the blood brain barrier unless they are lipophilic or fat soluble.

The issue of quantitative tumor burden as a determinant of response is another tumor characteristic of major importance. Tumor cells are killed according to the first order kinetics rule, and therefore a fixed percentage of residual tumor cells remain following any drug exposure or radiation insult. Thus, the larger the initial tumor volume, the larger the residual tumor volume following therapy. This simplistic explanation supports the rationale for employing adjuvant therapy in early cancer in which the host-tumor volume is small and is manifested as micrometastases.

Finally, the growth rate of a tumor correlates with responsiveness to therapy, although clinical studies have not definitively demonstrated this principle. The proposition that tumors of a rapid evolution are sensitive to therapy is rationalized on the basis that a major proportion of the tumor cells are in cycle and actively metabolizing, and are therefore more sensitive to the toxic effects of drugs or radiation. The most rapidly growing tumors clinically are Burkitt's lymphoma, oat cell carcinoma, leukemias and testicular cancers, all of which are exquisitely responsive to therapy. Slow growing tumors, such as sarcomas, colon cancer, and renal cell cancer, similarly appear less sensitive to therapy. This generality, however, has many exceptions clinically and the experimental data has not been definitively confirmed.

Measurements of the Effects of Therapy

The determination of the impact of therapy on malignant disease is a major area of interest and development in cancer research today. The distinction between objective tumor measurement, which involves specific quantitation of the tumor mass, and subjective tumor measurements, which may reflect qualitative host changes and secondary symptoms of the tumor as well as other factors not directly related to tumor mass, is crucial. Improvement in subjective symptoms may be secondary to supportive therapy or other changes in pathophysiology related to ancillary treatment and not to the tumor-specific therapy.

A listing of the objective measurable disease primers that one may employ in monitoring therapy are reviewed in Table 1.6. The most precisely measurable lesion is the discrete pulmonary nodule. Such lesions are discrete and definitively measurable to the point that growth rates of tumors may be predicted. The nonmeasurable lesions are collectively either a secondary manifestation of the metastatic tumor (e.g. biochemical abnormalities of liver function) or are categorically unmeasurable because of multifactorial influences on the tumor manifestation (e.g. skin ulceration may be complicated by local infection). In this latter example, healing may reflect infection control and not anti-tumor effect.

The semi-quantitative parameters are imprecise and almost empiric in that arbitrary criteria are defined and tested. Specific criteria for decrease in the size of a lesion do not necessarily indicate a precise tumor cell kill ratio. Response is most effectively judged if the tumor regresses completely. However,

Table 1.6
Objective Parameters of Cancer Serving as Measurable Lesions

Precisely Measurable	Pulmonary Nodule excluding Mediastinal Mass
	Skin Nodule
	Tumor Marker (Serologic HCG and AFP)
Semiquantitative	Hepatomegaly
	Liver Scan
	Pelvic Mass Lesions
Non-measurable	Pleural Effusion and Ascites
	Skin Ulceration
	Bone Lesions
	Subjective Symptoms
	Biochemical Measurements e.g. SGOT, Bilirubin

therapies generally bring about only a partial regression of the tumor and therefore a series of quantitative levels of response may be defined (Table 1.7). In the absence of measurable disease, the determination of effectiveness of therapy may be difficult. The most definitive criterion of measurement is improved survival, which is the final common goal for all clinical trials. If this is the only standard, however, other effective therapies, which are either transient or only effective for a small subset of patients, may be overlooked.

Table 1.7
Quantitative Measures of Response to Therapy (Particularly Chemotherapy)

Partial Response	50% decrease in product of maximum perpendicular diameters of any lesion lasting a minimum of 4 weeks and no new lesions
Improvement	25 to 50% decrease in product of perpendicular of maximum diameters
Stable Disease	less than 25% reduction in above "product"
No Response	more than 25% increase in product of maximum diameters or appearance of new lesions

---

8.0

Survival and the "5 Year Cure" Rule

The preceding discussion relates to patients with metastatic disease. In the case of adjuvant therapy—for patients with no residual or measurable disease—the method of measuring the impact of therapy on the disease is to monitor patients for survival and for intervals of time for relapse or recurrence. Large comparative trials employ either survival or the time from treatment initiation to relapse as primary determinants of the effectiveness of treatment. Sophisticated statistical analysis is used in these randomized clinical trials, including stratification of heterogeneous populations. In some studies, such as the Veterans Administration Cancer Studies, survival has been the primary measurement for patients with metastatic disease and for adjuvant trials. The 5 year survival rule is commonly used as an indicator of curability and basically states that patients surviving 5 years following the initial diagnosis without recurrence are unlikely to develop subsequent recurrence. For colon and lung cancer, patients surviving 5 years and remaining free of disease are indeed cured. However, this rule is not applicable for other tumors such as breast cancer and malignant melanoma in which late relapses are common. In fact, metastases in breast cancer may occur as long as 20 years following the primary therapy.

Experimental Clinical Trials

Cancer therapy today is not optimal for two reasons: (1) the majority of patients still die of their tumor and (2) there are a number of ongoing clinical therapeutic trials that are designed to identify more effective forms of treatment. It is important to understand what a clinical trial represents, what types of clinical trials may be performed, and what kinds of mechanics are involved in them.

A clinical trial is a type of biologic experiment, yet it differs from a laboratory experiment using test tubes or animals. The clinical experiment involves: (a) a heterogeneous population; (b) an informed subject; and (c) a substantial rationale for treatment that has been evaluated by pilot studies for tolerability and by animal studies for effectiveness.

Clinical trials are divided into two broad categories: pilot studies and cooperative group studies. A pilot study is generally a treatment regimen which is employed in a heterogeneous group of patients to establish the safety and potential effectiveness of the treatment plan. The second type, the cooperative group clinical trial, requires large patient populations to evaluate new therapeutic programs. The cooperative group is especially necessary in two cases: 1) when evaluating a treatment for a rare tumor, in which accrual sufficient to evaluate the disease is otherwise limited and 2) when testing a therapy that has only a modest impact on the tumor in which the statistical identification of a difference requires large patient accrual.

A clinical therapeutic trial often involves a comparison of treatment A versus treatment B. One of the treatment *arms*, as they are called, represents "standard" therapy for that stage and type of tumor while the alternative arm represents the "new" therapy. The experimental therapy, however, must be proven to be effective and tolerable (by being evaluated in a pilot study) before it can be used in a comparative study. An important aspect of a comparative clinical trial is the necessity for randomization or unbiased selection of the treatment modality. The purpose of randomization is to optimize (*not guarantee*) the chance for the results of the study to be statistically valid by insuring that patients entered on all arms of a study are biologically equivalent in spite of unknown variables which influence response. For known biologic determinants of response, randomization is complemented by stratification in which the patient groups are balanced and then randomized. If, for example, one knows that the age of the patient influences the response to a treatment then one insures that an equal number of patients of different ages are entered into the two arms.

Clearly, comparative clinical trials, randomization, and statistical sophistication are not warranted or even necessary for treatments which produce a major anti-tumor effect which is in contrast to all previous experience. For example, if the response rate to treatment for colon cancer is generally accepted to be 20% and the new treatment yields a response rate of 70%

then a comparative randomized trial is unnecessary. The use of historic controls or groups of patients which have been studied previously and are comparable in all important ways with the ongoing patient group, is scientifically valid.

Unfortunately, the cancer treatments employed to date have not had a dramatic impact on cancer. Consequently, comparative clinical trials are crucial to the advancement of our knowledge of cancer and to developing new therapeutic modalities.

References

Ackerman, L. V., and Regato, J. A., *Cancer: Diagnosis, Treatment, and Prognosis.* C. V. Mosby, St. Louis (1970).

Baserga, R. (ed.), *The Cell Cycle and Cancer.* M. Dekker, New York (1971).

Horton, J., and Hill, G. J. (eds.), *Clinical Oncology.* W. B. Saunders, Philadelphia, 1977.

Lawrence, Jr., W., and Terz, J. J. (eds.), *Cancer Management.* Grune & Stratton, New York (1977).

Staquet, M. J. (ed.), *Cancer Therapy: Prognostic Factors and Criteria of Response.* Raven Press, New York (1975).

2 Telling the Diagnosis

1.0

Introduction
One of the most important aspects of cancer care is to establish communication between medical care personnel (the health care team), the patient, and the patient's family. Physicians, nurses, dieticians, social workers, and the entire gamut of professional and support personnel all interact with the cancer patient. This communication system is crucial to the patient's concept of well-being and, if effective and appropriate for the patient, may make a meaningful contribution to the actual duration of life.

It is important that the health care team has an awareness of the setting and the process by which the patient reaches an understanding of his or her diagnosis. The actual encounter with the diagnosis can elicit a variety of reactions from patients. The patient who has not had any symptoms secondary to the cancer and who has not anticipated the disease should be confronted directly with the diagnosis. Alternatively, the patient who has been through a long period of chronic illness without an established specific diagnosis and an understanding of his or her symptoms may not be emotionally strong enough to handle a candid discussion of the diagnosis without experiencing severe depression. Most often, the patient will accept the diagnosis as a fatal perspective on his or her life. This patient requires input and frankness, and the health care team must develop the discussion at a slow, comfortable pace to help the patient deal with the personal ramifications of the diagnosis. This chapter will explore ways of approaching the patient with the diagnosis and will develop some of the techniques used in maintaining the patient's self-image, concept of well-being, and sense of hope.

2.0

Pros and Cons of Knowing the Diagnosis
The term cancer has ominous if not fatal connotations for most of us. As a result, we are often more comfortable in employing terms like tumor or neoplasm. In some cultures, the word cancer is excluded from the language. This denial is, of

course, extreme; but it is not uncommon in either Japan or France to interpret the diagnosis and to tell patients that they simply have a "blood" disease. In such settings, discussions of the diagnosis cannot include the important and relevant concepts of mortality and survival which are of primary concern. The patient who knows his or her diagnosis has as an immediate response, whether silent or verbalized: "How long do I have?"

The United States has moved from this restricted viewpoint because treatment techniques, which are widely available and often experimental, necessitate informed consent from the patient. Of greater importance, the patient must have an awareness of his or her disease in order to accept the substantial risks and discomfort involved in treating it. Furthermore, a knowledge of the diagnosis enables the patient to balance the discomfort of the treatment against the reality—that is, the fatality—of the disease. The patient will relate more appropriately to the therapists and will accept the recommended treatment more readily.

The reasons for not telling the diagnosis, or the cons, are most often imposed by the patient's family. The family may feel that the patient would not be able to "take it" and would die perhaps more quickly. This reaction often reveals the family's fear not only for the patient but for themselves. They are afraid that *they* might have cancer since someone so close to them does. In reality, it is quite rare that the knowledge of the diagnosis should lead to withdrawal or to a lack of a desire to live. In truth, confronting the diagnosis is an evolutionary process for the patient; it invariably results in a greater understanding and appreciation of life, family, and friends.

The pros for telling the diagnosis are compelling. First, the awareness allows patients to proceed with finalizing their affairs and planning their lives appropriately in terms of achieving goals. The college student who may have an estimated 2 years to live may not wish to complete the rigors of college life. Similarly, a young wife with an expected survival of not beyond 1 year may wish to defer adopting or bearing children in the face of an uncertain future. This situation is particularly poignant and may result in a decision to have a child, since raising children is often one of life's major goals or expectations. Aware of the imminence of death, the couple may choose to reinforce and perpetuate their relationship through a child. The child may be a burden for the remaining spouse but, obviously, this issue can only be confronted and dealt with in frank discussions. The importance of knowing the diagnosis in such discussions is obvious. Ignorance not only imposes constraints but also promotes impractical solutions.

A second major reason for presenting the diagnosis to the patient is to allow the patient to resolve interpersonal issues. Residual or even apparent major conflicts that the patient may have with family or friends may, unfortunately, be maintained up to the patient's death if the seriousness of the disease is not understood. Unresolved issues serve the surviving family members in an adverse fashion, leaving emotional scars of significant importance. The patient who knows his or her diagnosis becomes highly sensitive to the important aspects of life and is able to deal effectively with those minor irritations which create major conflicts between people who do not have to deal with their own mortality. Resolution of minor conflicts promotes the emotional stability of the patient and ensures the emotional support of the family.

Finally, it is easier for the physician and staff to effectively manage a patient who is fully aware of his or her diagnosis because they need not maintain a facade or engage in deception. Their energy can be more usefully employed in providing compassionate understanding and in promoting communication, which in the final analysis, is the most meaningful therapy the health care team can offer patients.

3.0 Cancer as a Chronic Disease

The apprehension which accompanies the diagnosis of cancer may be ameliorated at least in part by relating the clinical aspects of this disease to a more acceptable disease—one that is not associated with many of the fantasies and myths with which we imbue cancer. Cancer is a group of individual diseases, numbering more than 100 in type; each disease is specific and distinctive. Breast cancer is as different from lung cancer as coronary artery disease is from glomerulonephritis, a renal disease.

Making analogies among diseases is an important way of relating information to patients. Analogies help to relieve patients' anxiety when they learn the diagnosis for the first time and help them to understand the clinical aspects of their disease more clearly. In these situations, physicians should avoid the word cancer and refer to it instead as a neoplasm, tumor, and lesion.

Diabetes is a typical example of a disease that can be employed as an analogy. Like cancer, it is a chronic disease, requires constant monitoring, often involves continued treatment throughout the course of the disease with either pills or injections, may require intermittent hospitalizations, and compromises a variety of organ functions. Most patients can accept that diabetes is chronic and does not necessarily compromise their total life span. In fact, diabetes does affect longevity; but by replacing the cancer image with the diabetic image at least initially, one has simultaneously instilled the concept of disease and chronicity, but has diminished and diffused the terrifying mystery of visions of the future. It is important to remember that the analogy technique is employed only when broaching the diagnosis. With time the patient will need to be more realistic about his or her disease.

4.0 Techniques of Reassurance Therapy

Most patients have a premonition of their diagnosis before the physician actually discusses it in detail with them. This phenomenon may be related in part to subtle body language communication or to unrecognized clues in the demeanor of the staff as they interact with the patient. Nonetheless, the emotional impact upon the patient when actually hearing the word cancer, leukemia, or malignant melanoma is formidable and is equatable with the most terrifying of experiences.

There are two critical ingredients to the technique of telling the patient the diagnosis. The first is to fulfill the patient's need to be reassured that he or she will not be alone. The staff must constantly be sensitive to reassuring the patient that the health care team will always be available for his or her needs. The simple statement "I will always be here with you" can mitigate the sense of isolation the patients develops in response to hearing the diagnosis and may diffuse some initial hostility toward the world at large and the health care team in particular.

The second critical ingredient is to indicate to the patient that everything possible will be done to treat the cancer. This knowledge will allow patients to maintain a hope for survival and furthermore permits them to intellectualize about the therapy itself, thus permitting a smoother transition to acceptance of the diagnosis. The phrase, "We (I) will do everything possible to help," should be part of the initial contact and should be repeated and reinforced at intervals. Furthermore, the dialogue serves to solidify the bond between the patient and the staff, allowing the patient, at least within the crucial period of accepting the diagnosis, to develop a dependence on the staff which will serve as an avenue of communication.

Telling the patient the diagnosis also provides an opportunity to discuss the natural history of the disease, the extent of the disease, the form of the therapy that can be applied, and the effectiveness of that therapy. In essence, this approach becomes an important educational process. Ideally, it should also demythologize cancer fantasies, a process which will be explored more fully.

5.0
The Prognosis—The Primary Issue

When a patient receives the diagnosis of cancer, life's expectations are suddenly and devastatingly eliminated. The immediate question is "How much time do I have left?" But this question may be repressed because of the fear of the actual response by the staff and of the reality testing the question must naturally initiate in response. Physicians will often not discuss the patient's life expectations and the patient, unwilling to address this issue, will look for subtle, less precise, clues. For example, the patient with lung cancer whom the doctor does not advise to stop smoking quickly interprets this "response" as indicating a very short future. Or the patient may set up opportunities that may subtly reveal what the future may hold. For example, patients may ask whether it would be appropriate to plan for their daughter's wedding in the spring or fall. The doctor's response provides them with a relative indication of their expectation for the future. This

20

method of interaction between health personnel is often subject to misinterpretation and confusion, and a forthright discussion of the future and realistic explanations may obviate much of this playground tactic.

How does the oncology team approach the issue of prognosis with the patient? First and foremost, health personnel should never attempt to specifically state a time span expectation. Often patients will focus on that time (if they are told such a time) in a very precise and compulsive manner and will develop a physical and psychological withdrawal syndrome shortly before that date. One can usually not predict with even a modicum of accuracy the month or week of death of an individual patient. Patients who are clearly pre-terminal—that is, destined to die within 2 weeks—and who, from a clinical standpoint, have failed to respond to all therapy can have their demise predicted relatively accurately. Yet, even in these instances, although one may be confident of an expected duration of life, there is tremendous variability. In general, it should not be necessary to tell patients or their families the date of death as the end will be obvious to them all.

When discussing the future with the patient, it is best to avoid the use of the word "fatal." The discussion may involve a consideration of "incurability" which need not imply death and the expectation of living with the disease can be sustained. Incurability will still allow the patient to deal with the necessary rigors of treatment and does not eliminate hope. It is also advisable to avoid dealing with questions relating to how the patient may die. This morbid preoccupation with the mode of death is not uncommon and it is important that health personnel dissuade the patient from focusing on such issues for two reasons. First, this focus is destructive in that it may reinforce fantasies about death. Second, the intellectualization of death is an ineffectual defense mechanism which does not resolve the conflicts that patients must endure. It is more appropriate to work for a constructive resolution of the anxieties relative to the cancer. The principle stands that having established an understanding of the diagnosis and, in general terms, of the prognosis, the health care team must not focus on the details of the time and means of death.

6.0 **A Selection of Patient Reactions**
To this point, we have discussed ways of telling the diagnosis and of discussing the prognosis or future. The patient's ability to accept and interact with this new perspective on life and death depends to a large extent upon the ego strength and defense mechanisms the patient has employed in other stressful situations. Six categories of reactions may be established, although such separations are selective and may be considered artificial. The purpose of creating these categories is to help the physician and health care team identify, anticipate, and deal with patients' reactions. It is important to remember that these reactions often represent different components of a spectrum of response that may be experienced by each patient.

Denial: This mechanism of response is the most commonly observed and may be reflected in the patient's inability to inquire about his disease or his future. A more subtle instance of denial is that of the patient who continues planning long range activities involving major commitments when the need for intensive therapy is obvious to the patient, the family, and the health care team.

> *Patient T., a 29 year old war veteran, had an amputation for a fibrosarcoma of the left arm and persisted in completing his engineering degree with plans for entering law school and becoming a patent attorney. One year following the amputation he developed metastases in the lungs and refused drug treatment over a 3 month period during which time the tumors grew dramatically. When finally consenting to therapy, he referred constantly to his goal to be a patent attorney and graduated from engineering school while receiving chemotherapy and having developed brain metastases. Up to the time he developed progressive coma, he had plans to enter law school in the fall.*

There is no question that the future was a crucially important life and ego support concept for this patient. In such cases, it is the physician's responsibility to establish a realistic but not overly pessimistic outlook on the future for the patient. The patient must still feel that life is worth living, and that certain goals can still be accomplished or he or she may enter into a state of withdrawal and inertia. Thus, in the example just cited, the physician did not remind the patient relentlessly that his objectives could not be achieved. Rather, the physician interfaced extensively with the family (in the patient's absence) to ensure that they had a realistic appraisal of the future and would not entertain any false hopes.

Covert Anger: The patient who responds to his diagnosis with covert blatant hostility is generally one who has been oppressed or downtrodden for whatever reasons throughout his or her life and is in many ways unhappy. Another type that responds with hostility is the high powered executive who receives the diagnosis in mid career and is bothered by the interruption in his or her ascendancy and life's ambitions. The hostility may be directed at the physicians and their "incompetence," the nursing staff, or members of the family.

> *Y. is a 34 year old mother of three children who was a 3 pack a day smoker and developed an epidermoid carcinoma of the lung with a secondary SVC (Superior Vena Cava) syndrome. Shortly after completion of radiotherapy and resolution of the SVC syndrome, she developed an extensive retroperitoneal tumor. The patient and her husband were incredulous that the tumor recurred. His response to the diagnosis was tearful and she was unable to understand the situation at all. She repeatedly assailed her doctors with the fact that an x-ray, supposedly taken only 4 months previously, had revealed that her condition was normal. In*

spite of her intelligence and success in business, she was totally unable to accept that the same tumor had spread. The patient would treat the nurses and physicians as servants, asking them to perform personal services and frequently asking them to wait for her to complete some function, such as talking on the phone or shaving her legs.

The anger in this instance is subtle and is paradoxical in that the patient acknowledges her need for medical help and yet continuously challenges the medical profession. Such anger may be maintained right up until the time of death, often with withdrawal from medical care toward the end. The anger at this stage is then directed at the family as a substitute for the health care team. The resolution of the anger depends in large measure upon compassionate family support and upon understanding as well as confrontation by the physician and health personnel. The medical staff should point out the issues to the patient in candid discussions.

Dependency: Regression to infantile behavior and manipulation contrasts with the behavior of the hostile patient. Dependent patients actually interact constantly with the health care team and are often child-like in asking for help or explanations about their disease. This regression is often counterproductive and requires strong interfacing on the part of the team to overcome it. Persistence of such behavior is a tremendous strain on the resources and capacity for tolerance of the team. Encouraging independent activities and discouraging infantile behavior will bear the fruit of an accelerated convalescence.

W., a 49 year old woman, had had diabetes for 25 years which was controlled with insulin. She later developed an ovarian carcinoma which was treated by surgical resection. Post-operatively she was advised to have chemotherapy. The woman reacted adversely not so much to treatment itself but to the concept of needing such treatment, which indicated to her that she had an incurable tumor. Actually, her tumor had been removed completely and the treatment was being employed in an adjuvant circumstance. The patient became increasingly infantile which generally elicited the sympathy of the health care team, but she was a tremendous strain on their energies, requiring constant attention and tedious periods of explanation. She required advice on hair coloring, bed positions, time for bowel movements, etc., all in compulsive detail.

The reaction of the health care team should be stern and paternalistic, insisting on independency and emphasizing the need to be active and functional outside of the disease. By ignoring the patient's minute and menial concerns, the health team succeeded in making her overcome her dependency on them.

Withdrawal: Total withdrawal is an uncommon and paradoxical reaction. The patient often feels insecure and manifests this feeling as hostility. This hostility only results in

23

greater withdrawal. In order to suppress anger, the patient draws inward even more and thereby rejects emotional support from others. This reaction is unusual initially but may evolve after a long period of increasingly distressing symptoms and ineffectual treatment plans. It is impossible to deal with a patient's withdrawal without bringing an intermediary into the family. Psychosocial therapy is imperative and often helpful in diffusing anger and in allowing the patient's needs to be recognized. This type of psychological support is especially important for patients who have had their disease for a long time.

F. was a 65 year old man who had been through the Depression, the wars, and had been totally independent and without family for his adult life. He developed a crippling and devastating carcinoma of the anus and rectum, necessitating in sequence a colostomy, followed by pelvis radiation. He developed secondary sepsis and ulceration of the perineum and peri-rectal area. The patient was initially affable, hopeful, and oriented toward returning to work. With increasing lack of success of the therapeutic modalities, the patient became less communicative with occasional bursts of anger and impatience. Over the last 3 weeks of life, the patient assumed a fetal position in bed and refused food or attempts to engage in conversation with the health care team. He died not having spoken for 7 days, although he was awake and alert and responded to sounds.

This case is an extreme example of withdrawal in the hospital setting. Hospitals often augment patients' reactions because the patients are isolated from the comfort of their usual surroundings. Withdrawal, however, may occur in other settings and especially after a long series of hopeful therapeutic trials have failed.

Bravado: The patient who denies all symptoms and accepts all therapy with complete commitment is actually indulging in a form of withdrawal. In this instance the patient insists upon working, is more often a male than a female, and refuses to adjust his or her life style. The inability of such patients to recognize their needs is harmful. To insist upon pushing their strengths to the maximum is detrimental to helping them deal with their disease and symptoms. Many patients will endure overwhelming pain and discomfort to maintain the facade of health in a totally unnecessary context. In addition, this behavior pattern does not give the family the opportunity to relate to the patient in a protective fashion and thus to develop their own grief reactions. Direct confrontation is the only means of dealing with such bravado and such action is often rebuffed. The health care team should engage in slowly paced conversations with such patients to detect potential symptoms, to demonstrate concern, and to make them recognize that weakness is an acceptable part of being sick.

Patient M. is a 62 year old high-powered real estate saleswoman who had a large ulcerating carcinoma of the breast.

Prior to that development, she had a large mass lesion of her breast for approximately 8 years. During that period, she refused to acknowledge this major local discomfort and what it represented to her. She did not seek treatment and insisted upon working. Since she had let it go for 1 or 2 years, she believed that nothing could be done. She and her husband endured local ulcerative sepsis and a constant odor. It was only after she had achieved semi-retirement and had lived with the intolerability of the odor over some period of time, that she finally consented to medical care.

In this instance, the patient had a fear of medical care and refused to accept her need for medical assistance. Once having accepted care, however, she was enthusiastic and totally committed to whatever was prescribed. Mastectomy was followed by an immediate restitution to her normal life style without the need of convalescence.

The Realistic Patient: This category is an unusual one and may be characterized by two types of individuals. First, elderly executives, who have led full lives and have passed through life's passages successfully, will often accept the diagnosis and resolve issues within themselves and with their families with minimal trauma. Such patients are generally well adjusted to life and its exigencies. Many have already achieved their major goals and ambitions, and are now secure and dignified. In contrast to these patients, elderly executives who are still trying to fulfill goals—such as to become presidents of their companies—do not cope well with their diagnoses and often become hostile and intolerant because of their helplessness and inability to control their disease.

The second type of individuals in the realistic category are religious, often orthodox, individuals and those within the clergy. Nuns will often accept death as the joining with Christ in matrimony. The belief and commitment of the religious function as a major sustenance throughout the transient pre-terminal period; their ability to accept the diagnosis and the imminence of death is remarkable.

Here is an example of the first type:

O. was a 70 year old successful, retired banker who had been perfectly well throughout his life when he suddenly developed multiple bruises and a sensation of weakness with fever. He was found to have acute myelogenous leukemia and during the period of his diagnostic workup he insisted to the health care team that he wished to be advised immediately when his diagnosis was known. When the physician initiated the discussion of his diagnosis, the patient forthrightly requested detailed information about leukemia and the treatment alternatives. He expressed the desire to submit to a course of chemotherapy which would be maximal and if no response was obtained, then he wished to return to his home in the country where he could receive blood transfusions as necessary. This therapeutic plan was agreed to by the health care team and the patient did respond and lived without maintenance chemotherapy for approximately 10 months with his family at home.

This unusual patient was content with his achievements in life and had no fear of death. He had probably resolved the issues of life or had tired of the struggle to resolve them. He was contented and comfortable throughout his terminal phase and did not experience anxiety.

7.0

Cancer Myths and Fears

Patients' conceptions of their cancer contribute significantly to their ability to accept therapy and a new life style. It is crucially important that the physician and primary medical team ensure that patients have an accurate understanding of their disease and its ramifications. Distortions and myths about cancer will jeopardize the physical and emotional well-being of patients and will interfere with their therapy. An example of the impact of distortion is seen in the patient who develops a solitary metastatic lesion from a previous but remote cancer who may presume that he or she has only months to live because a relative (who may have had a totally different cancer) died 2 months after developing a metastatic lesion at the same organ site. With the expectation of dying within months, this patient may not go to see his physician for treatment.

Patient J. is a 60 year old woman who had a breast cancer surgically removed in 1965 and a second breast cancer removed in 1972. In the interim between the two tumors, she was perfectly healthy. In 1976 the patient developed local skin metastases on her chest. Believing that her surgeon could not do anything for her, she did not see him when her routine follow-up visit came around. Three months later she decided to see him simply to say goodbye since she had a warm affection for him. The surgeon placed the patient on a hormonal regimen and the lesions of the chest wall disappeared completely.

In this instance, the patient acted upon the false assumption that cancer is a fatal disease that can be treated only once. Not appreciating the fact that breast cancer is exquisitely sensitive to treatment, she was denying herself medical care. Fortunately, her bond with the physician encouraged her to interact with him again.

It is now apparent that patients do not communicate distortions and cancer myths to health personnel. The health care team may reverse this trend by creating a comfortable atmosphere that will encourage patients to speak openly. Professionals must not only be able to detect signs of misapprehension, however, but to anticipate them and to know how to respond. Queries should be dealt with frankly and responses should promote understanding, acceptance, and hope as much as possible. We will now discuss potential queries that patients may propose and will present appropriate responses by the health care team.

How did I get this?

The development of cancer always brings with it the question of the cause, and patients often feel that they may have done something specific to bring it on. They interpret cancer as a punishment for their sins and it becomes a "dirty disease," bearing a stigma like venereal disease. These attitudes are destructive and unnecessary. The health team must prevent the patient from creating an ever-expanding circle of self-doubt and self-blame.

The health care team's response to this query should educate the patient as comprehensively as possible. For example: "Unfortunately we do not yet know the specific cause of any of the multiple types of cancer. There are known carcinogenic chemicals such as the tars in cigarettes, but in addition there is an inherent susceptibility (sometimes genetic but not transmissible) of the host to develop a malignant transformation when the appropriate stimulus comes along. The stimulus may be a chemical, a virus or some other circumstance."

In spite of such enlightening reassurance, patients may persist in believing that some external or internal mechanism, such as evil thoughts, contributed to the development of cancer. In relation to the reaction types described earlier in this chapter, the hostile response to the diagnosis is often accompanied by the question, "What did I do to deserve this?" It is difficult for the health care team to deal with this question as it is merely rhetorical. The response in the mind of the patient is: "Nothing." The patient's illness is just another indication of how the world is "screwing me," or how "I am a perennial loser." The best method of handling these angry and paranoid attitudes and the verbalized or silent reaction is to recite the facts about cancer: 1) It is the second most common disease of importance (not, "the second most common killer"); 2) There are many different forms of treatment; and 3) It has causes which are largely unknown but general clues are developing.

Is cancer hereditary and will my children get it as well?

This is a common misconception and should be approached directly. Often patients will not wish to ask the question because they are so fearful of the answer. Cancer is rarely hereditary. Although one can speculate that cancer develops in a susceptible individual and that susceptibility may be familial and transmissible, it is *not* at all true that cancer is a genetically determined disease. It may appear to the layman that cancer is hereditary in that many patients will have relatives with cancer, but this fact is more a reflection of the commonness of the disease than the familial transmission of it.

When establishing a patient's medical history, however, it is important to elicit the family history. A patient may already have some understanding of cancer since other family members may have had the disease. There are also some rare tumors which are definitely inherited and examination of family

27

members may be indicated. Most importantly, the family history will help the health care team to understand the patient's reactions.

Will my husband (or wife) be likely to contract this tumor?

An illustrative case will highlight the poignancy of this question. A woman who developed carcinoma of the cervix received radiation therapy. Two years later she developed local recurrence high in the vaginal vault. She had at the inception of the diagnosis discontinued all sexual relations with her husband, and had withdrawn from her family which had been previously a happy and prosperous one. Her unvoiced concern was that the tumor could be transmitted to her husband whom she loved so dearly.

The converse clinical situation may also be observed. A woman who develops carcinoma of the cervix may blame her husband because of the "statistical" association of cervical cancer with a herpes virus present in the secretions from the male penis. The patient who is aware of this theory from the newspapers or lay publications may be unduly sensitive and is certainly inadequately informed. Nonetheless, she may irrationally attribute the development of her cancer to her husband's promiscuity.

It is important in this context as well as in others that a clear statement be made as to the communicability or lack of it in the case of cancer. Cancer is not at all contagious and the transmissibility issue should be broached directly. If the patient doesn't inquire, a common syndrome often related to the patient's desire to avoid talking about such issues, then the issue should be raised by the health care team. Direct confrontation can only affect the patient's and the family's life style in a beneficial way even if only on an unconscious level, as observed in the cited examples.

How will I die?

Along with the issue of "How long do I have?" is the concern of how the patient will die. Most patients have the misconception that death is equivalent to suffering, and that cancer in particular is a disease associated with suffering and pain. They often attempt to understand the actual physical process of dying.

The physician or health personnel must discuss the question frankly when the patient raises the issue of death patterns. However, the issue of how one dies should not be discussed unless: (1) the patient directly confronts the health care team with the question, and (2) it is evident that death is imminent in weeks. It is important to establish the motive behind the question: is it an intellectual one or does it reflect a deep-seated fear? In the completely asymptomatic patient who has no life style compromise in either the immediate or semi-immediate future, it is not recommended to discuss the circumstances of dying. Rather, the patient's morbid preoccupation should be assuaged by suggesting that such concerns are irrational in the face of obviously excellent health. With the patient who has a critical short time interval remaining or who is symptomatic in a major way, the question could be addressed more directly and specifically. In addition, compassionate reassurance must be the overriding concern of the interviewer, and sensitive listening in a quiet and comfortable environment is essential. It is reassuring for patients to know that they will: (a) not have pain, (b) not be a burden to their families or to the health care team, and (c) will be totally cognizant and aware up to the point where they will enter a coma or protracted sleep.

The simplest explanation to the persistent query "How will I die?" is that the tumor may interrupt the function of an essential organ as the liver, the lung, even the brain, or possibly the kidneys, resulting in a progressively comatose state. It is extremely unusual for patients to die with recalcitrant pain or to endure a long period of pain prior to dying. In a study by Osler, less than 5% of patients dying of cancer actually had pain as part of their symptom complex in the time immediately prior to death. This type of information is sometimes critically important for patients to understand and incorporate into their expectations but as indicated previously may represent a mechanism of intellectualization about the cancer. Presenting the reality that death is often a process of progressive sleep is often sufficient. However, patients may wish to know if death will be painful, if they will choke to death or lose control of their excretory function; in short, they wonder if they will become uncomfortable, dependent, and embarrassed. In such instances, the health care team should reassure the patient and emphasize the potential for reversing these effects. What these patients are asking and revealing is that they would rather die while they have all their faculties.

8.0   **The Role of the Family**
One of the most crucial responsibilities of the health care team is to relate effectively with the family and to develop the family's capability to care for the patient. The team, therefore, should not only counsel the patient but should address itself to the stress reactions of the family and help them develop supporting structures to deal with the disease and the future. The family who insists that the patient could not stand to know the diagnosis should be carefully informed of the pros and cons of knowing the diagnosis as well as of the importance of honesty. The protective reaction of the family is actually one of self-protection and should be uncovered for all the family members.

The use of group meetings of families, excluding the patient but involving other families in the same situation, is often a helpful technique. Families, with members at similar stages of the disease process, meet to share their experiences and to develop better ways of relating to the patient and his or her coming death. Such groups are perhaps best maintained at no more than three or four families with no more than two or three members per family. The role of the psychologist, social worker, psychiatrist, health care physician, or nurse in such a group would not only be to provide medical information, but more importantly to help the group evolve toward a commiserating congregation of mutual support. Cancer is a disease which affects the entire family and the educational process must involve them as well as the patient.

9.0 **Conclusions**

The questions and concerns raised in this chapter are often not broached by the patient and it is encumbent upon the team to either elicit or initiate discussion about them. The patient will be looking for "receptive" clues to be able to talk and will be responsive to any clues the team may generate. Again, the major concern of the health care team should be to deal with the apprehensions and myths of cancer early on in order to create a favorable atmosphere for treatment, and to ensure maintenance of a productive life style for the time the patient has remaining.

3   Common Queries on Cancer Therapy

1.0

Introduction

The myths and fantasies that surround cancer as a disease also encircle its treatments. Chemotherapy and radiation therapy are as mysterious and awesome to patients as the disease itself, and the thought of receiving them often provokes extreme reactions from fear to panic. It is important to educate patients and to coordinate the details of their therapy with their life style and personal needs. This planning can evolve in response to patient inquisitiveness or can be initiated by nurses and physicians who may broach various areas spontaneously during the course of treatment. But the initial preparation is crucial and the health care team must guard against their own assumptions and casual acceptance of the treatment modality and relate the basics of the treatment to each patient in a simple way.

Health personnel, involved in cancer and its intricacies, often forget that patients look upon many of the procedures, tests, and other aspects of cancer management in total ignorance and fear. Imagine a patient trying to understand why the doctor takes a blood count when the tumor is in the bone. To this end, this chapter is devoted to answering some of the questions that patients undergoing therapy often ask.

2.0

A Synopsis of Common Queries Related to Specific Treatments

I.   *What are the Current Forms of Treatment for Cancer?*
     Surgery, radiation therapy, chemotherapy, and immunotherapy are the most common types of treatment for cancer. They are used individually or in combination with one another.

II. *What is Chemotherapy?*
     Chemotherapy is the use of drugs for the treatment of many diseases, including cancer.

A. How is chemotherapy given?
   1. Orally—by mouth
   2. Intramuscularly—a "shot" into a muscle
   3. Intravenously—an injection into a vein under the skin
      a) Intravenous treatment may take an extended period of time, usually under a half hour but occasionally an hour or more.

B. Is chemotherapy painful?
   Usually not, except for the initial needle insertion. Occasionally its administration causes discomfort.

C. How does chemotherapy work?
   The drug gets into the bloodstream directly by intravenous injection or indirectly by absorption through the stomach or the tissues. When the drug reaches the tumor cell, it acts like an antibiotic fighting germ cells during an infection. The drug prevents the tumor cells from growing and causes cells to die. The successful destruction of the tumor depends upon several factors including the type of tumor and the drug. While the drugs affect both normal and tumor cells, normal cells can regenerate themselves, returning to their normal state. In an effort to insure the safest and most effective treatment, the patient's blood is tested regularly to monitor the effects of drugs on normal cells.

D. How often is chemotherapy given and for how long must one receive this treatment?
   The length of time and frequency of chemotherapy depends upon the type of cancer the patient has, the drugs necessary for treatment, how long it takes the patient's body to respond to these drugs, and how well he or she tolerates them. Treatment schedules vary widely. The duration of treatment may range from a few months to several.

E. What are the possible side effects of chemotherapy?
   Some of the drugs used produce side effects, also termed "toxicity." Although all patients do not experience all side effects, most patients do experience some of them. The degree of severity varies. In most patients, such side effects are usually reversible. Some potential side effects are:
   1. Gastrointestinal effects
      The patient may experience nausea and vomiting. In some cases, this reaction may be severe. Antinausea pills and suppositories may be prescribed by the physician to alleviate the problem. Various sedatives might also be used to help the patient during this period. Other gastrointestinal effects might include diarrhea, constipation, and ulcera-

tion of the lining of the mouth (mucositis). Instruction in mouth care and medications can help relieve pain from mouth ulceration. These latter effects clear up within a fairly short time.

2. Muscle and nerve effects

   Weakness and lethargy in moving muscles is occasionally observed and may be related to anemia, but can also be a direct effect of some of the drugs on the muscles themselves. In addition, the effect on the nerves may result in a tingling or burning sensation in the hands and feet. This feeling is similar to having one's hands "fall asleep" and may be associated with some clumsiness of movement. These temporary effects are reversible in most cases.

3. Bone marrow effects

   The bone marrow is where your body manufactures white blood cells which combat infection, red blood cells which prevent anemia and allow the blood to bring oxygen to all the tissues, and platelets which help clot blood and promote healing of breaks in the skin. As mentioned earlier, chemotherapy acts to prevent cells from duplicating and sometimes acts on normal cells. Since bone marrow cells duplicate rapidly in order to maintain the blood counts, these cells are particularly sensitive to chemotherapy. These potential changes in the blood are temporary. Transfusions of white cells, platelets, or red cells are available to help counteract these effects.

   a) A blood count may be taken at regular intervals so that the health care team can watch for these effects:

      (1) a decrease in white blood count which may result in the development of infection;

      (2) a decrease in platelet count which may result in the development of bruising, a skin rash of small blood blisters under the skin, or actual bleeding internally from the stomach or kidney;

      (3) a decrease in red blood cells which might cause anemia and result in shortness of breath, weakness, and fatigue.

   b) If the patient develops:

      (1) a low white count, the patient must notify the doctor of any temperature elevation or any new symptoms;

      (2) a low platelet count, he or she must avoid the sun, alcohol, aspirin, or any other medication unless approved by the physician.

4. Hair effects

   The hair follicles of the scalp and beard as well as

eyebrows, eyelashes, armpits, and pubic and leg hair are rapidly growing cells and are sensitive to some forms of chemotherapy, but not all. Partial loss of scalp and body hair commonly accompanies treatment, with occasional complete loss. Hairpieces are tax deductible medical expenses and are sometimes covered by insurance. All body hair will return completely when drugs are discontinued and occasionally may return partially during treatment.

5. Testicular and ovarian effects

Women who are still having menstrual periods commonly develop irregular periods or experience cessation of menstrual flow during chemotherapy. In women close to the menopausal age, hot flashes may develop.

Conception during treatment is unlikely, but contraception should be continued. In men chemotherapy often results in reduction of sperm count and viability of the sperm, causing sterility. Treatment does not interfere with the ability to continue normal sexual relations. Patients have experienced resumption of normal testicular and ovarian function, and have conceived normal children following discontinuation of treatment. However, it is possible for the fertility of the male to be permanently curtailed by some treatment programs. For treatment programs with such potential, sperm may be frozen and stored for future artificial insemination.

6. Skin effects

There are a variety of rashes which may develop, although only infrequently. Localized or generalized rashes are usually red and sometimes associated with itching. Occasionally, the skin may peel, but blister formation practically never occurs. The drugs may also irritate the veins and result in coloration of the venous system along the pathway of the vein.

III. *What is Immunotherapy?*

Immunotherapy (or immune therapy) is a form of cancer treatment which attempts to stimulate the natural body defenses or immune system. Its goal is to activate certain types of white blood cells and proteins that make up the immune system to fight off cancer cells.

A. How is immunotherapy given?

Immunotherapy may be given by scarification. This method is similar to the way smallpox vaccine is given. Immunotherapy may also be given in small injections just underneath the skin. At regular intervals the patient's blood will be tested to detect changes in the immune system. Skin tests will be done to determine the effects of treatment. The specific treatment is explained to the patient in detail by the doctor and the nurse.

B. Is immunotherapy painful?
The injection or scratch (scarification) itself may cause some discomfort at the time of administration. Most patients feel this irritation is minimal.

C. How often is immune therapy given and for how long must one receive this treatment?
The duration of treatment will be determined by the patient's blood tests and periodic physical examinations. Some patients will receive therapy according to specific protocols or treatment plans which will be thoroughly explained to them.

D. What are the possible side effects of immunotherapy?

1. Side effects will vary with specific types of treatment, but in general patients experience some local itching, or inflammation at the site of the injection or scarification. Also, many have flu-like symptoms the evening of the treatment, including mild fever, chills, and muscle aches. These symptoms are much like those felt after a vaccination.
2. If immunotherapy is given by injections under the skin, the site may weep for a short period of time. The patient will be instructed in the care of these areas.
3. For some types of immunotherapy, reversible liver changes may occur. The patient's blood will be checked routinely for these symptoms so that treatment may be changed accordingly, if necessary.
4. Additional side effects or problems will be discussed with the patient by the nurse and doctor.

IV. *How does Radiation Therapy Work?*
Radiation therapy, x-ray treatment, or irradiation treatment is the use of the energy produced by x-rays to interrupt the growth of the tumor cell or to destroy it. This energy is like the energy rays that come from the sun, which are x-rays as well as ultra-violet rays. The ray acts like a drug by interrupting an important chemical process in the tumor cell, which results in its self-destruction.

A. How is Radiation Treatment Given?
The radiation treatment is very much like a routine x-ray in that the patient is simply sitting or lying in

a room for approximately 1 minute while the machine is on. There is absolutely no pain, discomfort or even awareness of any sensation. By and large the treatments are administered for 5 days a week from 2 to 6 weeks depending upon the area where the tumor is found in the body and the kind of tumor present. Since radiation treatment is almost always administered to a localized area, the doctors will outline a "port" or area within which radiation will pass through the patient's body. This outline, made by small tattoo marks under the skin, is a guide to focus the treatment ray and therefore protect normal, non-tumor-bearing tissues from radiation damage.

B. Are There Side Effects of Radiation Therapy?
There may be some side effects which are similar to those of chemotherapy. When the abdomen is radiated the result may be some nausea, vomiting, and decreased appetite. Such an effect may be experienced even when an area outside the abdomen is radiated, related probably in part to a type of radiation sickness. Other types of side effects are possible but they are temporary and of relatively little concern. Such effects include: a reddening of the skin, hair loss, and possibly some generalized weakness.

C. Can Radiation Therapy Cause Cancer?
Radiation has definitely caused cancer in experimental animals and in humans. However, the dosage being employed for patients is directed at eradicating an already present tumor and is not associated with the induction of cancer. This dosage would effectively eliminate any of the standard forms of radiation-induced cancer.

3.0   **Questions Related to Understanding the Rationale for Treatment**

I.   *Why is Treatment Recommended?*
Past experience has taught doctors that in some cases even when the tumor has been completely removed by surgery, there is a possibility that it will recur. "Adjuvant therapy" is treatment directed at preventing such a recurrence. Treatment may also be recommended for patients when complete removal of the tumor has not been possible or when the tumor has already spread. In this case the goal is to shrink the tumor(s).

II.   *How is my Specific Treatment Determined?*
Many factors will be considered by the doctor in determining specific treatments. The most important factors are: the kind of cancer; the sensitivity of the tumor to specific drugs or other modalities of treatment such as

surgery, radiotherapy, or immunotherapy; and the areas of the body affected by the tumor.

The patient's treatment program may be called a "protocol." This therapy is a predetermined treatment plan for groups of patients with similar types of tumors. Some protocols call for "randomization," the unbiased assignment to one of several treatment groups or "arms" (see question XI). Such comparative studies are called "clinical trials" and are designed to determine the best possible treatment. This procedure often involves testing a new therapy program against the "standard" treatment to see which is more effective.

All treatment plans are reviewed by a scientific committee to insure their validity. These programs are also reviewed by the Human Protection Committee, composed of professional and lay persons from various hospitals and the community. The committee's primary concern is to protect patients' rights.

It is important that patients are not placed in a clinical trial without their knowledge and full consent. The consent form which patients are asked to sign before beginning treatment is designed to insure that they understand the potential goals of the treatment and its possible side effects.

Patients may ask to be withdrawn from any protocol at any time without jeopardizing their care in any way. Doctors should explain in detail the pros and cons of a recommended treatment as well as alternative forms of therapy.

III. *How Can the Effect of Treatment on the Tumor be Measured?*
Physical examination, x-rays, and various laboratory tests allow the health care team to evaluate the tumor's response to treatment.

IV. *What if the Chosen Treatment does not Work?*
With few exceptions, there are many alternative treatment programs, one of which may result in regression of the tumor even after another has failed to adequately control it.

V. *Why is Chemotherapy Needed?*
Drug treatment destroys the tumor or prevents it from returning. In some situations, although the surgeon or the radiotherapist has completely removed the tumor, it may return or spread to other areas of the body. In this case, drug treatment is extra insurance and will, hopefully, prevent the tumor from returning. If a tumor could not be removed completely or if it has spread to other areas, drug treatment is intended to cause regression of the tumor. Drug therapy, unlike radiation therapy, will be distributed through the blood stream to all parts of

the body and if effective will work wherever the tumor may be.

VI. *Why is Radiation Therapy Needed?*
Radiation treatment is used either to prevent a tumor from recurring in the local area from which it was removed, or to cause a tumor in a localized area to regress. If a patient has symptoms induced by a tumor growing in a specific area, the radiation therapy (if effective) will cause regression of the tumor and will thereby alleviate the symptoms.

VII. *If One Does Not Get the Side Effects of the Treatment is the Drug or Radiation Still Working?*
There is absolutely no correlation between toxicities or side effects and the destruction of a tumor. Although most patients experience secondary symptoms, some experience none. A patient should not feel that simply because he or she is sick, the drug is working more effectively.

VIII. *Why are Blood Tests Needed Periodically?*
Blood tests are intended to act as a guide for the doctor to regulate the dosage of a patient's drug or radiation treatment. Both the drugs and the radiation may affect the blood forming tissues in the body and make the red blood cells, white blood cells, or platelets decrease in number. Therefore, blood tests do not necessarily reflect the status of the tumor but are important in helping adjust the treatment.

IX. *Is this Treatment Experimental?*
Therapy can be "experimental" to a very limited extent. Although a patient may receive a drug that is standard or that has been available for a long time, a doctor may use it differently to improve its capacity to control the tumor.
In some instances, a patient may be treated with a totally new drug which has not been used previously. The doctor will carefully explain the reasons for choosing this type of treatment. The decision to accept this form of treatment, however, is completely voluntary and requires the consent as well as the understanding of the patient. In general, a treatment is "experimental" because the health care team will be carefully evaluating the effectiveness of the drugs in relation to alternative treatments. The goal is always to improve on known effective treatments.

X. *What is the Purpose of Signing a Consent Form?*
The consent form is designed to insure that patients understand the goals of the treatment and the possible side effects. It does not obligate patients to start or to complete the treatment and patients may withdraw from the specific treatment at any time and consider alternatives. The consent does not relinquish any of a patient's rights

40

or the health team's responsibilities to insure total safety and the correctness of treatment. In addition, all programs are reviewed by a Human Rights Committee, composed of people from the community who are not physicians, to guarantee that the rights of patients are not denied.

XI. *What is Randomization and will it Guarantee the Best Treatment?*
Randomization is simply an unbiased method of choosing therapy when there is more than one choice and perhaps multiple choices, all of which are judged to be relatively equal clinically and experimentally. The "best" treatment can only be established by a direct comparison and randomization enables one to definitively establish which treatment is superior. Randomization, therefore, is a process that ensures that the "arms" of a protocol are equal with regard to those features of a disease that influence the prognosis. The singular difference between the arms is then the treatment. The actual choice of treatment is arbitrary.

XII. *Can Drug Treatment or Radiation Make the Cancer Grow Faster?*
Definitely not. This very common misconception developed because in the past, only patients with extremely advanced tumors received chemotherapy, and it often appeared as if many died soon after. Their deaths, however, were related to the advanced state of the tumor. There is no evidence whatsoever that drugs will make a tumor grow faster.

4.0   **Questions Related to Life Style Adjustments**

I. *Will the Patient have to be Hospitalized for Therapy?*
Most chemotherapy programs can be accomplished in the outpatient department. For some patients or treatment programs, however, a short period of hospitalization may be necessary.

II. *Will the Patient be able to Continue Working?*
The degree to which people are able to work and perform their customary activities varies. Many patients on chemotherapy tire more easily during treatment. Clinic personnel (social workers and nurses) are available to assist patients in planning the extent to which they can carry on regular activities and occupations.

III. *What About Taking Other Pills or Drugs During Treatment?*
A patient should consult the doctor before taking any other pills. Patients should bring all the pills which they have been taking recently to the doctor's office. The doc-

tor can then evaluate the situation. Table 3.1 presents a list of those drugs which may possibly interfere with or in some way affect chemotherapy.

IV. *Should the Patient's Diet be Restricted in Any Way?*
Generally no. Patients receiving chemotherapy should eat a light breakfast on the day of their treatments. They might also be advised to increase their fluid intake before, during, and following their treatment by about 2 to 4 glasses of water per day. If patients have any problems with diet, or if they are not eating very much, they should not hesitate to inform the staff. The clinic dietician can then give them instructions.

V. *Will the Patient be able to Have Dental Work Done While on Treatment?*
Generally yes for routine check-ups. However, a patient should consult the doctor before visiting the dentist.

Table 3.1
Drugs Which May Affect the Patient's Treatment

The Doctor Should be Notified if the Patient Takes Any One or More of These Drugs:

Aspirin
Darvon
Barbiturates (Seconal, Nembutal)
Sleeping pills
Cough medicines, including Robitussin
Blood pressure pills
Anticonvulsant (seizure) pills
Diuretics—"water pills"
Hormone pills (birth control pills)
Antibiotics
Anticoagulants or blood medications
Tranquilizers or "nerve pills"
Arthritis or bursitis pills

VI. *When Should a Patient Call the Doctor?*
A. Patients should not hesitate to call the doctor if they have any symptoms that worry them, especially in the following circumstances:
1. Fever (temperature of 100° or over)
2. Development of any rash
3. Any kind of bleeding which persists
4. Any pain of unusual intensity or distribution, including headache
5. Shortness of breath
6. Any side effect that provokes worry

VII. *May the Patient Have Sexual Relations and How Often?*
This question is often not voiced and should be raised by health care personnel, particularly for young persons who

may be sexually active. By and large there are no restrictions unless genital structures are compromised by the tumor or the treatment—for example, there may be a rash. The treatments do not lead to impotence but the illness itself may contribute to disinterest.

VIII. *Can the Patient Travel, Especially by Airplane?*
Travel is rarely restricted. The exertion required, however, may limit the patient's capability or endurance.

Section II

Review of Therapeutic Modalities

4

Chemotherapy: General Concepts

1.0

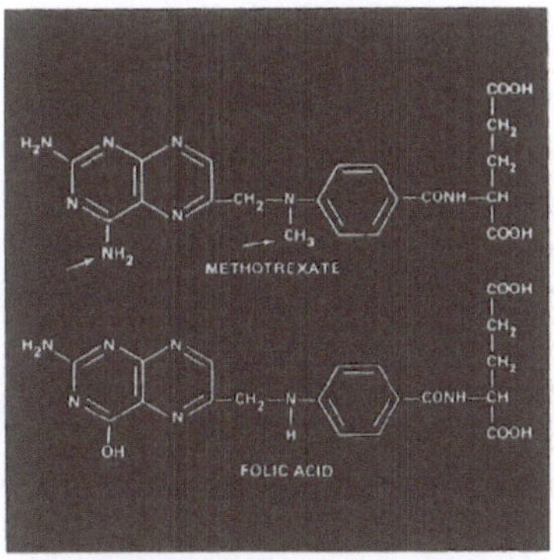

Figure 4.1
The structural formula of Methotrexate. This antimetabolite tumor drug differs from folic acid, the normal physiologic vitamin supplement, in only two areas, as indicated by the arrows.

Introduction
There are approximately 80 anti-tumor drugs employed to treat cancer patients today. Many of these drugs were discovered by virtue of animal tumor screening programs, some by rational chemical design, and others by astute clinical observation. An example of a drug developed by rational design is 5-Fluorouracil. Heidelberger created this drug after analyzing many human and animal tumors and discovering that malignant cell growth (in contrast to normal cell growth) was heavily dependent upon uracil as an essential component of DNA. He reasoned that distorting the uracil molecule by attaching a halogen, fluorine, may result in a DNA which is incompatible with tumor cell viability. Methotrexate is an example of a drug discovered by clinical observation. Dr. Sidney Farber was the first to detect that leukemic children developed progressive disease when placed on vitamins which included folic acid. The folic acid antagonist Methotrexate (see Figure 4.1) was just then in the process of development and Farber demonstrated in 1948 that an analogue drug of Methotrexate could induce remission in acute leukemia. The great majority of agents, however, have been incorporated into clinical use because of empiric drug screening. The plant alkaloid derivatives, for example, were identified as active anti-tumor agents in the process of screening any and all plant components of all species for such activity.

The collective anti-tumor drugs may be classified into categories according to the mechanism of cell interaction, the chemical derivation, or the source and composition of the drug. A common drug classification scheme identifies five drug classes: antimetabolites, alkylating agents, antibiotics, natural products, and a miscellaneous category. There are multiple drugs within each category, some of which are closely related analogues, but each of these drugs has special pharmacologic or biologic properties and therefore has specific indications.

Drug Development

The identification of active cancer chemotherapeutic drugs is a major goal of cancer institutions such as the National Cancer Institute and the Regional Cancer Centers. For the most part cytotoxic drugs are developed by the pharmaceutical industry and by biochemical investigators screening in sequence series of compounds in animal tumor models. Drugs found to be effective in eradicating animal tumors then undergo a specific process of drug evaluation in which toxicologic study in animals eventually leads to clinical therapeutic trials in man. In general, the animal tumors employed for testing are transplanted tumors which are developed or induced in one animal and are then transplanted and maintained in a large series of animals of the same species. Table 4.1 lists a selection of the common animal tumors used in drug development studies.

Table 4.1
Animal Tumor Systems Employed in Testing New Cancer Drugs

Tumor	Animal
L 1210 Leukemia	Mouse
P 388	Mouse
Sarcoma 180	Mouse
B 16 Melanoma	Mouse
Lewis Lung Carcinoma	Mouse
Walker 256 Sarcoma	Rat
AKR Lymphoma	Mouse

The next step in the developmental sequence is to determine the adverse effects that drugs may produce on the normal dog, monkey, mouse, and guinea pig systems. On the basis of these preclinical toxicology evaluations, a dose and schedule for the drug can be established. It is important to note that the adverse effects of drugs may not be detected in preclinical testing; nonetheless, a guide to potential side effects is created. Up to this point, the potential drug has been developed by individual scientists in basic biologic research.

The clinical use of such drugs in humans similarly progresses through a predetermined scheme of development which is designed to establish the toxicologic limitations of the drug and to identify its specific therapeutic activity in tumors. In Table 4.2 the developmental phases of investigative clinical trials are described. Phase I trials establish the best route, schedule, and dose of drug in humans, as well as the patterns of toxicity. In addition, pharmacology studies of the new drug determine the metabolism and organ distribution *in vivo*. The goal of Phase II clinical studies is to specifically evaluate anti-tumor effects of the drug in specific types of signal tumors (such as breast, lung, lymphoma, and leukemia) which are representative of the gamut of common responsive and resistant tumors. In Phase III trials a comparison of the

Table 4.2
Drug Development and Clinical Trials

Phase I: Clinical pharmacology and toxicology

Establish maximum tolerated dose at schedule(s) tested.

Establish toxicity parameters and determine if toxicity is predictable, treatable, and/or reversible.

Pharmacologic evaluation.

Anti-tumor activity not necessarily required.

General total number of patients required 20 to 30.

Phase II: Screening and clinical activity

Treat 20 to 30 patients with measurable disease in each of a range of "signal" tumor types.

General total number of patients entered >100.

Phase III: Trial for establishing recommendations for general use

Controlled randomized and comparative clinical trials in a specific tumor.

General number of patients needed depends on activity of standard therapy.

new drug to "standard therapy" is undertaken generally in a randomized trial. In this context, the new drug therapy must be equivalent to or superior to standard therapy in the nonrandomized Phase II trials.

Drug Classes

The cancer chemotherapeutic drugs are separated into five classes, arbitrarily based on the mechanism of tumor cell kill. In addition to the mechanistic classification, agents may be classified according to the timing of the anti-tumor activity relative to the growth cycle of the tumor cell (see section 4.0). Within each category at least five separate and distinctive drugs may be identified, some of which are analogues, by which is meant a minor chemical alteration results in a major biologic difference. For example, the two anthracycline derivatives, adriamycin and daunomycin, differ by a single hydroxyl group (Figure 4.2) and the two compounds have a distinctively different spectrum of activity. The drug representative of each category of activity as well as those drugs with unusual pharmacologic or therapeutic implications are reviewed by category.

Alkylating Agents

The alkylating agents are the oldest of the anti-tumor drugs. These drugs (Table 4.3) are extremely reactive compounds which interact with the chemical composition of DNA and create disruption of the genetic material. Nitrogen mustard is the original alkylating drug. In general, alkylating agents have a very short half-life in the plasma, interacting with cells im-

3.0

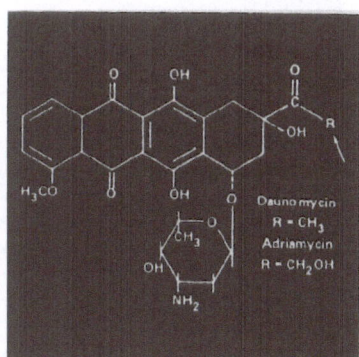

Figure 4.2
The structural formula of the anthracycline antibiotics, Adriamycin and Daunomycin, differ at only a single site represented by hydroxalation of the methyl group.

3.1

47

Table 4.3
Alkylating Agents in Cancer Chemotherapy

	Mechanism of Action	Major Toxicity
Mustards		
nitrogen mustard chlorambucil cyclophosphamide melphalan	Alkylating of macromolecules, cross-linking of DNA	Bone marrow depression
Methanesulfonates		
busulfan	Same	Myelosuppression
Nitrosoureas		
BCNU, CCNU streptozotocin	Alkylation plus additional mechanisms	Delayed myelosuppression Islet cell and renal damage

mediately after injection. Cyclophosphamide is a unique drug in that it requires metabolic activation in the liver prior to affecting the tumor. Therefore, patients with liver disease may be more sensitive to cyclophosphamide, and drugs which affect the activation mechanisms may also distort therapy. The nitrosoureas are unique compounds because they are lipid soluble and cross the blood brain barrier. The effect of this drug on the bone marrow is often delayed, requiring it to be given at intermittent intervals with a 4 to 6 week period between drug exposures. Cumulative marrow suppression is common with these agents, although some analogies are only minimally myelosuppressive, such as streptozotocin and chlorozotocin.

Unusual but not uncommon side effects for the alkylating agents include the hemorrhagic-cystitis associated with cyclophosphamide (which may be mitigated by high fluid intake), and the bulsulfan induced lethargy syndrome of pigmentation and asthenia. Generally, the alkylating agents in contrast to the antimetabolites induce infertility and amenorrhea, and the potential for carcinogenesis and mutagenesis is well established.

The tumors most effectively treated with this class of drugs include: ovary and breast cancer, multiple myeloma, and lymphoma. Other categories of tumor may be treated with alkylating drugs successfully, but not as consistently as those cited.

3.2 **Antibiotics**

The antibiotic compounds consist of substances derived from bacterial colonies which are highly noxious to normal human cells and, more importantly, to tumor cells. The antibiotics employed as anti-tumor compounds are more potent than those, such as penicillin, used to combat infections. Consequently, normal cells have less tolerance for anti-tumor drugs.

48

The mechanism of the tumor cell kill for each of the antibiotics in this class is distinctly different; for example, mitomycin C is an antibiotic which may function as an alkylating agent; mithramycin affects tumors by binding intracellular heavy metals. These drugs generally have a short half-life because of rapid urinary secretion and hepatic degradation. Renal and liver functions may affect the pharmacology of the drugs. The anthracycline antibiotics, including Actinomycin D, are actively secreted in the biliary system and pathologic obstruction of the bile ducts leads to prolonged plasma drug levels and accentuated toxicologic effect. The specific drugs within this class affect a variety of neoplastic disease but the anthracycline compounds have the broadest spectrum of activity (Table 4.4).

Table 4.4
Antibiotics in Cancer Chemotherapy

Antibiotics	Mechanism of Action	Major Toxicity	Tumor Affected
Actinomycin D	Binds to DNA and inhibits RNA production	Marrow GI	Wilm's Tumor Testicular
Anthracyclines (Daunorubicin, Adriamycin)	Damages DNA and inhibits RNA production	Marrow	Breast, Leukemia, Ovary, Sarcoma
Mithramycin	Binds Magnesium	Marrow Liver	Testicular
Bleomycin	Damages DNA	Skin Lungs	Epidermoid, Testicular
Mitomycin C	Alkylating drug	Marrow, Kidney	Gastric

3.3

Antimetabolites

The antimetabolites are a group of drugs are a complex group of drugs about which much biochemical information has developed (Table 4.5). These drugs directly affect DNA synthesis

Table 4.5
Antimetabolites in Cancer Chemotherapy

Drug	Mechanism of Action	Major Toxicity
MTX	Inhibits folate reductase, and secondarily, DNA synthesis	Marrow GI
MP and TG	Inhibits purine synthesis	Marrow GI
FU	Inhibits DNA synthesis, thymidilate synthetase	Marrow GI
ARA-C	Inhibits DNA synthesis via DNA polymerase	Marrow

49

by interrupting the essential production enzymes or by directly inhibiting the synthesis of pyrimidine or purine, both of which are necessary components of the DNA molecule. These drugs affect cycling or growing and metabolizing cells. Although the end result is similar to the alkylating agents (DNA interruption), the biomolecular effect is distinctive.

5-Fluorouracil (5-FU): The fluorinated antipyrimidine drugs (including FUDR) directly inhibit thymidylate synthetase and prevent formation of thymidine, an essential component of DNA. The drug is activated intracellularly by a kinase and may be incorporated directly into the DNA or RNA molecule, producing a "fraudulent" structure. Generally FU is administered intravenously either weekly or in intermittent courses for 5 day intervals. The drug is rapidly cleared from the circulation and may be administered by bolus, pulse injection, or by continuous intravenous infusion. Intra-arterial pump perfusion directly into a tumor site, such as the liver, has also been employed (see Chapter 14). FU is most often employed for gastrointestinal tumors but it has also been demonstrated to be effective for breast and ovarian tumors.

Methotrexate (MTX): The antifolic acid compounds directly inhibit dihydrofolate reductase, thus preventing donation of a methyl group. Methotrexate, the most common type of folic acid antagonist, is not metabolized in the body but is excreted rapidly by the kidneys with an extremely short plasma half-life. In clinical circumstances where renal function is marginal, methotrexate treatment is associated with prolonged blood levels, but, in addition, methotrexate may be directly toxic to the kidney, and may induce severe although transient renal failure. A new method of methotrexate administration referred to as "high dose methotrexate with citrovorum rescue" (HD MTX/CF) has been developed for experimental use and involves the administration of extremely high doses of the drug which would ordinarily be potentially lethal. By providing the antidote, citrovorum or folic acid, the normal host tissues are "rescued," while the tumor cells are incapable of absorbing the antidote, resulting in tumor cell death.

Specific precautions in the use of high dose methotrexate necessitate that the drug be employed in this fashion only in institutions experienced in its use. One major limiting feature of HD MTX use is renal failure. Alkalinization of the patient with Na HCO$_3$ is necessary to increase the solubility of methotrexate and to prevent renal failure. The leukemias, breast, head, and neck cancer are the most common tumors for which MTX is employed.

Cytosine Arabinoside (Ara C): This drug was identified in screening compounds derived from sea life. Originally found in the sea sponge, the drug has been synthesized and is the most important compound in the treatment of leukemia. The drug is administered by continuous infusion or frequent parenteral injection because it is degraded rapidly by the deaminases. These enzymes reduce the blood levels of the active principle. The major dose-limiting toxicity of Ara C is myelosuppression and gastrointestinal effects. The primary, if not exclusive, use of Ara C is for acute leukemia, and possibly lymphomas.

Antipurines: This group of drugs inhibits purine biosynthesis and, like the antipyrimidines, require intracellular activation. In general, these drugs (6MP and 6TG) are administered as oral tablets on a daily schedule and may produce a fraudulent DNA, as the mechanism of cell kill. Toxicity, other than myelosuppression, is not significant with the exception of occasional hepatic damage. The primary therapeutic role of the antipurines has been in the treatment of leukemia.

3.4

Natural Products and Miscellaneous Agents (Table 4.6)
The periwinkle alkaloids, vincristine and vinblastine, are derived from plants and are therefore considered natural products. They are unusual for their effect on the microtubular structure of the cell—particularly in nerve cells—which accounts for the neuropathy commonly observed with prolonged treatment. The drugs are excreted by the biliary tree, and hepatic obstruction generally leads to prolonged blood levels and increased toxicity.

There is a broad range of miscellaneous drugs. They are generally synthetic compounds and the mechanisms of tumor cell kill are varied. Many of the drugs do not have general applicability, but rather are used for treatment of specific tumors almost exclusively—for example, DTIC is used for melanoma, procarbazine for Hodgkin's disease.

Table 4.6
Natural Products and Miscellaneous Agents in Cancer Chemotherapy

Drug	Mechanism of Action	Major Toxicity
Hydroxyurea	Inhibits ribonucleotide reductase	Myelosuppression
Imidazole Carboxamide	Alkylation and other complex mechanisms	GI Limited marrow suppression
Procarbazine	Multiple actions including H_2O_2 production	Myelosuppression
L-Asparaginase	Degrades essential asparagine amino acid	Hepatic, pancreatic, anti-anabolic
O'P'DDD	Necrosis of adrenal cells	GI, CNS
Vinblastine	Damages microtubular function and arrests cells in metaphase	Leukopenia
Vincristine	Same	Neurotoxicity

51

Hormones

Hormone therapy is a unique form of chemotherapy in that the cytotoxic mechanism is basically unknown. More intriguingly, this therapy is specific for the tumor cells and therefore has a minimal effect on the host except for those physiologic effects resulting from an alteration in the hormonal milieu. There are basically four hormone classes which make up the exogenous hormone therapies: estrogens, progesterones, androgens, and corticosteroids. Two additional agents, one of which is not yet available in the United States, are cyproterone, an anti-androgen, and tamoxifen, and anti-estrogen. These two drugs have been employed in the treatment of prostatic and breast cancer in Europe. There are several types of cancer which commonly respond to hormone treatment: endometrial cancer is treated with progesterone; breast cancer with any one of the four preparations; renal cell carcinoma primarily with progesterones but also with androgens; and prostatic cancer exclusively with estrogen preparations. Table 4.7 presents a summary of the formulations of the various hormone preparations. Two points with regard to hormone therapy are important. First, the response to such drugs is often major and it is not uncommon for the tumor to regress completely. Sequential hormone manipulations may result in repeated responses; for example, breast tumors may respond to estrogen withdrawal and alternative hormonal procedures after an initial response to an estrogen administration. Secondly, additive hormone therapy, particularly for breast cancer, may result in exacerbation or progression of the disease with stimulation. The stimulation may be detected by increased bone pain, hypercalcemia, or actual tumor growth. Such stimulatory responses are important in predicting the potential tumor response to the deletion of endogenous hormones by operative procedures. Operative forms of hormonal management are oophorectomy, adrenalectomy, hypophysectomy, and orchidectomy. The rationale for such procedures in breast cancer is

Table 4.7
Hormone Preparation

Class	Agent	Dose/Route
Progesterone	Medroxyprogesterone	100–200 mg/d po 200–600 mg twice weekly i.m.
Estrogens	Diethyl Stilbestrol	15 mg/d po
Androgens	Fluoxymethyl Testosterone ethanvate	10–20 mg/d po 600–1200 mg/wk i.m.
Corticosteroids	Prednisone	60–100 mg/d po
Anti-Estrogen	Tamoxifen	20 mg BID po
Anti-Androgen	Cyproterone	Not Available

depletion of endogenous hormone compounds. Although the ovary is a major source of estrogen, the adrenal gland is a secondary source. Adrenalectomy removes the second source of estrogens and hypophysectomy removes the stimulus to the ovaries and the adrenal glands (LH and ACTH respectively), thus reducing the normal estrogen level to zero. Similarly, for prostatic cancer, the androgen is produced primarily in the testis but also in the adrenal glands. Therefore, ablative procedures on the adrenal or pituitary gland may deplete the physiologic tumor supporting hormone.

Use of hormone management is basically empiric but some generalities may be stated. A woman's menopausal status affects the type of therapy she receives; premenopausal patients are treated with estrogen depletion and post-menopausal patients with additive estrogen. In addition to this general guideline, clinical responsiveness may be predicted on the basis of disease free intervals and metastatic sites of disease. For example, patients in whom the interval from diagnosis of the primary tumor to diagnosis of metastases exceeds 2 years are more likely to respond to therapy than patients with intervals of less than 2 years. Furthermore, patients with metastatic disease in the liver or in a lymphangitic pattern in the lung are less likely to respond to therapy and have a shorter life expectancy than patients with bone or skin metastases.

The estrogen receptor test performed on tumor tissue is an objective parameter that correlates with the clinical parameters cited above and is similarly predictive of response to hormonal management.

4.0 Mechanism of Drug Effect and Drug Resistance

Anti-tumor drugs kill cells by inducing specific biochemical disruptions which cannot be compensated for by the tumor cell. This loss of an essential biochemical capability results in tumor cell death. The specific biochemical events most commonly interrupted are DNA or RNA synthesis. In addition, protein synthesis may be primarily affected, or cell wall alterations may be created by the drugs. The arrest of the biochemical process is followed either by immediate death of the cell (intermitotic) or by the traversal of the cell through one or two mitotic cycles before death. If the cell contains the capability for repair of the biochemical or biophysical insult then the tumor cell will survive.

The growth cycle pattern of the cell is important to therapy in that drugs affect cells at different phases of growth (Figure 4.3). The phases are schematic conceptions of cell kinetics and may serve as a guide for clinical therapeutic management. Tumor cells in a resting phase of growth (Go) may be totally insensitive to therapy. Slow growing tumors presumably have a high proportion of cells in Go and would therefore not be responsive to antimetabolites. Although this therapeutic premise is generally true, it is not uniform. For example, the converse clinical circumstance would suggest that rapidly growing tumors would be most sensitive to antimetabolites.

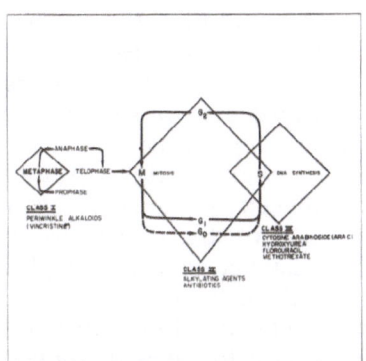

Figure 4.3
The various components of the cell growth cycle are indicated in relationship to the effect of the various categories of cytotoxic drugs. DNA synthesis (S) and protein synthesis (G_1 and G_2) are active metabolizing phases. The mitotic phase (M) is relatively brief. The majority of the tumor cells in a tumor lesion, however, are in the G_0 or resting phase.

Clinically the most rapidly growing tumors are most effectively treated by alkylating agents which are effective against Go cells as well as cycling cells.

The lack of selectivity of drugs for tumor cells results in the destruction of normal cells, known clinically as toxicity. Balancing the anti-tumor effect against the normal cell effect defines the therapeutic index of a drug or treatment regimen. A high therapeutic index is most desirable. A major component of the therapeutic index is the drug dosage. A dose response relationship exists for many if not all of the anti-tumor drugs such that the higher the dose, the more tumor cells killed. However, the patient's tolerance limits the ability to deliver a dose which would definitively eradicate the tumor. At present, the dose necessary to kill all the tumor cells would kill the host (patient) before eradicating the tumor.

A current understanding of tumor quantitation is that clinical detection of a tumor by either x-ray or palpable skin nodule represents a minimum of 10^{12} cells or more than a billion cells. A fixed proportion of the total tumor cell mass is killed by drugs with each drug exposure, the so-called first order kinetic mechanism. It is estimated that the maximum effect that regimens can achieve with complete clinical regression of a tumor is a thousand-fold reduction to 10^9 cells—at this point the tumor is not detectable by clinical means. Thus, one billion cells may remain and the patient has no evidence of clinical disease. It is at this juncture that immune therapy may be considered. Immune therapy is relatively free of host effects, may be tumor-specific, and is theoretically most effective against small tumor burden. It has been estimated that the maximum tumor burden for which immune therapy may be effective is only 10^5 cells. The hiatus between 10^5 and 10^9 tumor cells is an enormous gap to be considered in developing therapy for cancer and the problem is compounded by the fact that tumor cells may develop resistance to effective drugs.

The development of drug resistance by tumor cells is a complex problem and of major practical importance. How do tumors previously responding to therapy become resistant to drug treatment? The mechanism of resistance in tumor cells has only recently become appreciated. Tumor cells are not unlike bacteria in that resistance to drug therapy, like resistance to antibiotics, is often induced with selection of the hardy drug insensitive strain while the more sensitive strain is effectively killed off. Other mechanisms of resistance involve complex biochemical processes. For example, the cells may develop an enzyme for degradation of the chemotherapy drug intracellularly. Intracellular inactivation of the drug by a new tumor cell enzyme system or the development of cell wall changes, precluding transport of the drug into the cell, are additional theoretic mechanisms of tumor resistance. In general, once tumor cells have demonstrated resistance to therapy, continued treatment is not warranted with that particular drug. Instead non-cross resistant drugs should be employed. Cross-resistant drugs are most often interrelated in terms of both the mechanism of

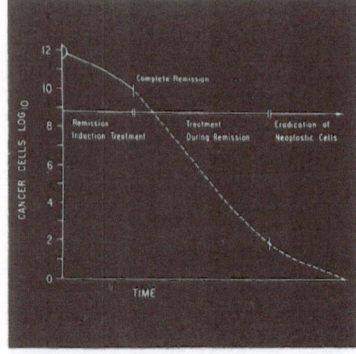

Figure 4.4
A diagramatic representation of tumor regression relative to quantitative tumor burden and the therapeutic strategy. Following remission induction, a persistent and significant tumor burden necessitates consolidation treatment and maintenance therapy.

tumor cell kill and chemical structure. Thus, all alkylating agents are cross-resistant so that changing treatment from cyclophosphamide to an alternative alkylating agent, such as chlorambucil, is not therapeutically beneficial.

5.0

Host Effects of Drug Therapy

As stated previously, anti-tumor drugs affect not only the tumor cell but normal host tissues as well. This lack of selectivity is reflected in the myriad side effects or toxicities that result from the use of such drugs. Normal tissues which demonstrate the greatest proliferative activity—that is, have the highest turnover rate or cell replacement—are the tissues most susceptible to side effects. The three tissues in the body in this category are the bone marrow myelopoietic cells, the hair follicle cells, and the gastrointestinal tract cells. Cytotoxic therapy, for example, frequently causes bone marrow suppression with pancytopenia, alopecia of the scalp as well as other body hairs, and gastrointestinal effects such as diarrhea. There are, in addition, side effects which are not predicted on the basis of growth rate of the tissue. Table 4.8 summarizes the side effects of cancer chemotherapeutic agents and focuses on therapy or intervention procedures. Many of the common side effects will be expanded in subsequent chapters and only the unique and specific toxicities will be discussed here.

Dermatologic Effects: Cytotoxic drugs are basically foreign substances and may produce allergic skin reactions. In addition to allergic reactions, there are a multiplicity of dose related effects on the skin which are characteristic for some drugs and related in part to localization of the drug within the epidermis. Two reactions are sufficiently unique and characteristic to warrant discussion. The first is the typical papulopustular lesion that develops secondary to Actinomycin D or Methotrexate. The skin reaction is rather typical and develops approximately 7 days following exposure to the drug but may occur as early as 3 days. This process may be generalized or, particularly in the case of Actinomycin D, localized to areas of previous radiation. The "recall" of prior radiation exposure by drug therapy is illustrated in Figure 4.5. The rash generally recedes within 7 days and may be associated with secondary infection related to the pustular inflammatory response. Methotrexate can produce a similar rash and may also induce recall of prior radiation.

A second unique rash is the desquamation syndrome which develops as a consequence of acute interruption of the epidermal layer of the skin. The superficial epidermis scales or sheds off and with denudation of the superficial layer, a "scaled skin" appearance is observed. This skin is typical for Methotrexate, particularly when administered at high doses or administered in association with prolonged high blood levels. A more localized reaction may develop in association with Bleomycin in which desquamation of the skin, particularly of the hands and around the ears, is common (Figure 4.6).

Table 4.8
Common Side Effects of Cancer Chemotherapeutic Drugs

System Involved	Mechanism of Effect	Action and Intervention
I. Gastrointestinal Tract (Chapter 8)		
A. Nausea and vomiting	Central effect on emesis center of the thalamus	Phenothiazine drugs (see Chapter 8)
B. Diarrhea	Local mucosal irritation secondary to the direct effect on the bowel	1. Prescribe anti-diarrhea drugs as necessary 2. Evaluate diet and advise low roughage foods 3. Increase fluids to prevent dehydration
C. Anorexia	1. Central effect on hunger center in thalamus 2. Secondary reaction to the nausea and vomiting 3. Nonspecific metabolic abnormalities	1. Determine dietary habits and plan foods with major appeal 2. Small, frequent feedings 3. Refer patient to a dietician
D. Taste distortion (dysgusia)	Local mucosal irritation	No known treatment
E. Constipation	Hypomotility of the bowel due to direct effect on the nerve supply (vincristine)	1. Use of cathartics and enemas as necessary 2. Dietary guidance
II. Skin and Mucosal Effects		
A. Dermatitis	1. Erythroderma; pustular or macular rash; desquamation and epidermolysis, all secondary to the direct effect on the skin 2. Allergic reaction mediated through immune mechanisms	1. A. Reduce dosage or withdraw drug B. Emollients such as AlphaKeri 2. Withdraw drug
B. Stomatitis (mouth ulcerations)	Direct effect on cells with inhibition of duplication	1. Local analgesics (also systemic) 2. Debridement (by water spray) 3. Mouth rinses of equal parts of hydrogen peroxide, water, and Cepacol mouthwash, q.2–3 hrs. 4. Mouthwash with neomycin and polymixin for bacteriostatic action 5. Suction as needed for excess secretion 6. Vaseline to lips 7. Soft foods as tolerated; feeding tubes may be necessary
C. Alopecia	Direct effect on cells	See text
D. Skin pigmentation	Unknown	See text
E. Nail changes	1. Onycholysis 2. Pigmentation lines secondary to growth arrest	Reassurance
F. Perianal and vaginal ulcerations	Direct effect on mucosa	Topical anesthetics, treat secondary infection (see Chapter 9)
III. Hematologic Problems (Chapter 5)		

A. Bone marrow suppression
   Leukopenia
   Anemia
   Thrombocytopenia
	Direct effect on the marrow stem cells arresting production of peripheral blood elements	1. Frequent blood counts (usually weekly)
		2. Teach patients to watch for signs of fatigue, weakness, fever, etc.
		3. Transfusions when necessary (PbC, WBC, platelets)
		4. No intramuscular injections

System Involved	Mechanism of Effect	Action and Intervention
B. Immune suppression	Decrease in T and B lymphocytes and antibody production	Nonspecific immune stimulation (see Chapter 6)
IV. Urinary Tract		
A. Hemorrhagic cystitis	Local bladder irritation	1. Increase fluid intake 2. Notify patient of symptoms to look for: hematuria, burning
B. Colored urine	Direct excretion of the drug adriamycin (red): methotrexate (yellow)	Monitor BUN and creatinine clearance; decrease dose or withdraw drug if BUN rises
V. Hepatic Effects	Fibrosis or hepatitis-like effect, cause unknown	Withdraw drug
VI. Reproduction System		
A. Amenorrhea	Involution of ovarian function and estrogen secretion	1. Reassurance—menses usually return to normal after discontinuing drug 2. Continue contraception
B. Impaired spermatogenesis	Involution of testicular function	1. Reassurance—no relationship to impotence 2. Pretreatment sperm banking
C. Premature menopause	Involution of ovarian function and estrogen secretion	1. Hot flashes and other symptoms may be treated with hormone therapy if the tumor treated is *not* breast cancer 2. Tranquilizers
VII. Neurologic Effects	Ataxia, paresthesias, paralyses by inhibition of microtubular substance in nerves (vincristine)	1. Modify dose or discontinue drug 2. Supply supportive care: walker, cane 3. Analgesics for pain 4. Reassurance that these effects are temporary and will usually disappear in 4–6 weeks
VIII. Miscellaneous Symptoms		
A. Fever, chills	Probable allergic reaction (bleomycin, DTIC)	1. Evaluate for incidental infection 2. Treat with antihistamines or corticosteroids
B. General malaise	Unknown	
C. Local vein and skin effects	Direct irritant of endothelium or subcutaneous tissues	Debridement; local steroids (see Chapter 14)

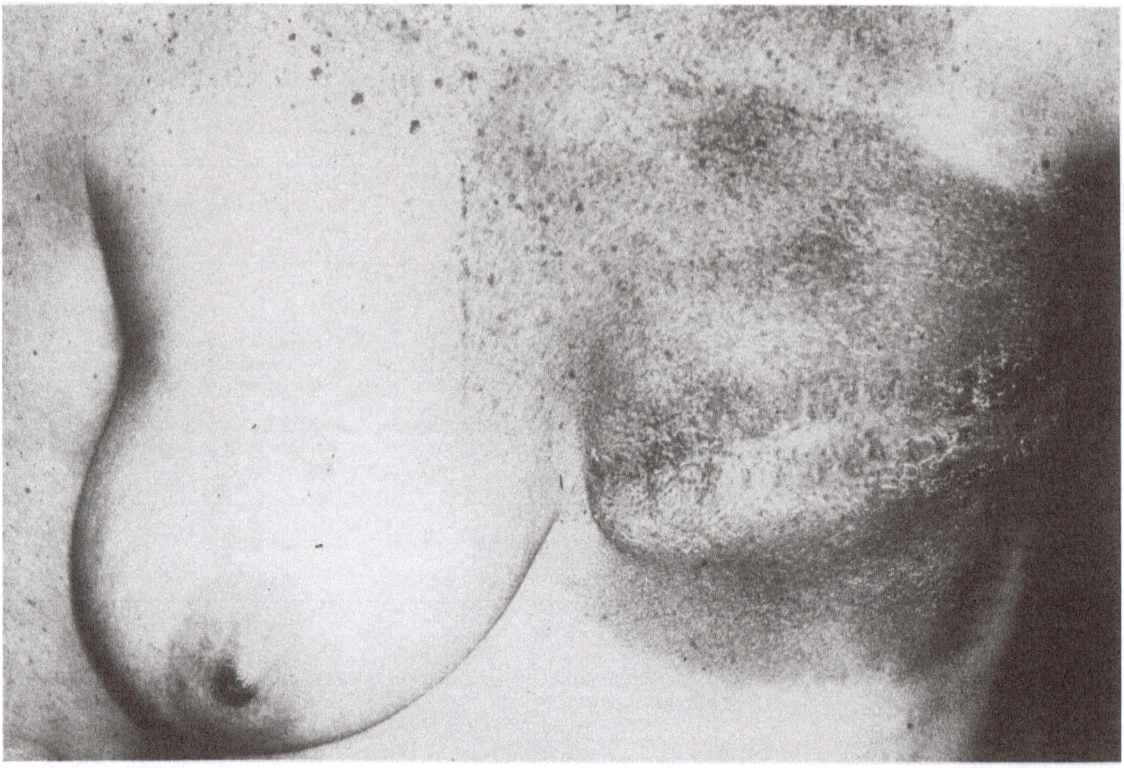

![Figure 4.5 photograph]

Figure 4.5
Following mastectomy, this patient received prophylactic radiotherapy to the chest wall in association with cytotoxic chemotherapy. The accentuated pigmentation indicates the areas of radiation application.

Another type of skin reaction is pigment deposition in the local cutaneous layers. Such an effect occurs particularly in patients treated with Adriamycin and is often localized to the mucosal surfaces. It can also occur on pressure areas. This effect is more pronounced in patients who have pigmented skin such as blacks and the Mediterranean ethnic groups. The pigmentary effect has generally not been readily reversible. The dark lines developing in the nail beds are related to the growth arrest and not to direct pigmentation. Another type of pigmentary skin effect is related to the disorganized use of iron as a consequence of secondary disruption of the synthesis of red blood cells.

Reproductive Effects: Many of the cytotoxic drugs induce a profound effect on the hormone secretions of the ovary and testicle. As a consequence, estrogen depletion and secondary menopause may develop; in men decreased sperm production is common although impotence rarely if ever result. The menopausal symptoms will recede if drug therapy is abbreviated and the dose of therapy sub-maximal. The most critical determinant of the development of drug induced menopause is the physiologic age of the patient. Patients who are approaching menopause have a marginal reserve of estrogen secretion and upon drug exposure may enter a permanent menopause. Untoward symptoms related to the traditional flushing of estrogen withdrawal may be obviated by the addition of hormones if the tumor being treated is not hormonally sensitive—that is, is not breast or uterine cancer.

Cytotoxic drugs have a major effect on fertility. These drugs

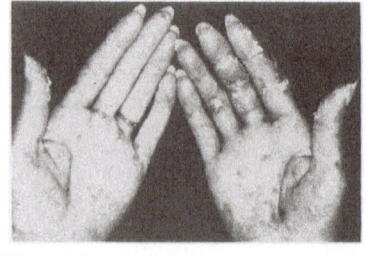

Figure 4.6
The palmar surfaces are initially covered with a macular papular rash after which desquamation of the lesions develops. These lesions, secondary to bleomycin drug treatment, heal completely within 7 to 10 days.

often function as contraceptives in the sense that they alter the menstrual cycle and produce anovulatory circumstances. The drugs are also effective abortifacients and abortion is not uncommon in pregnant women, particularly in the early stages of implantation, who are receiving treatment for cancer. This effect of the drugs is, however, unpredictable and standard forms of contraception should not be abandoned during therapy. With regard to male infertility, the presence of cancer may induce a relative sperm depletion state. If indeed a patient has potentially curable cancer and is being treated with anti-tumor drugs and particularly alkylating agents, the predictable development of azoospermia should be discussed with the patient. The azoospermia produced by drugs may be permanent without recovery even after discontinuation of therapy. Sperm banking facilities are available and patients may wish to take advantage of them so that they may potentially father children in the future. It should be emphasized, however, that specimens obtained from patients with cancer are often depleted of sperm because of the disease itself. In short, optimal specimens may not be available.

Finally, there is a real potential that cytotoxic drugs are mutagenic. Distortion of the sperm or ovum DNA content and subsequent chromosomal abnormalities with secondary congenital anomalies can occur. Studies to date have not yet established that mutagenicity is a major complication of drug therapy, but the experimental data is conclusive.

Cardiac Effects: The effect of cytotoxic drugs on the heart is generally minimal and if present is rarely significant clinically. The exception to this generality is the known cardiotoxic effects of the anthracycline antibiotics. Daunomycin and Adriamycin are capable of producing a cardiomyopathy which is dose related and which generally occurs above the dose of 500 mg/M². The principal manifestations of cardiac toxicity are progressive congestive heart failure and arrythmias. Such effects may be acute and the mortality may be as high as 10%. In the presence of radiation to the chest and heart, there may be a synergistic or potentiating effect of Adriamycin. Thus the maximum tolerated dose of Adriamycin is less than 500 mg/M² in such patients.

Pulmonary Effects: The drugs which affect the lungs are busulfan, an alkylating drug; methotrexate, an antimetabolite; and bleomycin, an antibiotic. The alkylating agent produces a chronic pulmonary interstitial fibrosis after long periods of drug exposure. Methotrexate has been associated in isolated cases with an acute alveolar pneumonitis. The most common and well established pulmonary toxicity effect of drug therapy is that caused by bleomycin. This antibiotic commonly produces an interstitial pneumonitis that may be acute in which case mortality approaches 80%, or chronic with interstitial fibrosis which resolves slowly with drug withdrawal. The latter clinical picture develops in a dose related fashion beyond 200 mg/M² cumulative. The acute interstitial process is idiosyncratic and may be seen with as little a dose as 45 mg/M² (Figure 4.7).

59

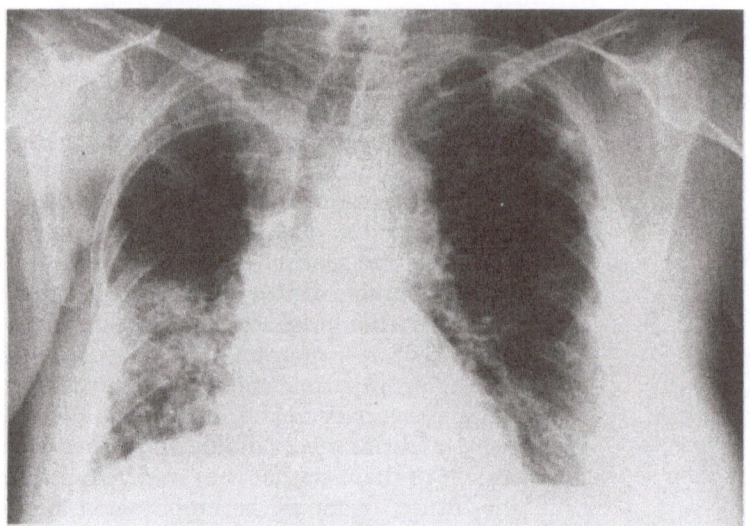

Figure 4.7
"Bleomycin lung" may manifest itself as an acute interstitial pneumonitis with fever and dyspnea. This patient progressed rapidly toward death.

6.0

Combination Chemotherapy

Multiple drug therapy represents one of the major advances in the application of chemotherapy in the past 10 years. The fact that drugs as single agents are effective in only a modicum of patients and lack specific selectivity for the tumor represents a major impediment to the development of effective therapeutic programs. The concept that non-cross resistant drugs with varying toxicities may be used to affect different parts of the cell growth cycle heralded a new era in the development of effective drug regimens. Combination chemotherapy programs of 2, 3, and even up to 7 drugs employed concomitantly have resulted in an augmentation of the capability to induce tumor regression. This augmentation may be more than additive and actually be synergistic with some drug combinations.

Combination chemotherapy principles are best illustrated with the combination of nitrogen mustard (Mustargen), Oncovin or Vincristine, prednisone, and procarbazine, commonly known as MOPP therapy for Hodgkin's disease. This combination was developed because the components of the combination of 4 drugs are non-cross resistant and affect different parts of the cell cycle. Most importantly, there is very minimal overlap of the toxicities of the various agents. Individually, each drug is known to be active in Hodgkin's disease and may induce responses in up to 20% of patients. When employed in a combination, the response to therapy approaches 80%. A primary principle of combination chemotherapy is the importance of employing drugs which have definitive activity against the tumor even if such activity is only marginal.

In addition to multiple drugs administered simultaneously, another form of combination chemotherapy is the sequential application of intermittent courses of alternative drugs following recovery from the myelosuppression induced by the previous drug combination. Intermittent sequential therapy allows for immunologic reconstitution and a delay of tumor cell resistance to drugs.

60

Today combination chemotherapy is the cornerstone of drug treatment, and single agent therapy is reserved for development of new drugs in Phase I and Phase II studies. Multiple drug therapy is employed as standard treatment for metastatic breast cancer, testicular cancer, soft tissue sarcomas, and the gamut of hematologic malignancies.

7.0

Response to Chemotherapeutic Agents

There is a general impression that chemotherapy has little if any effect on tumors. This misconception developed partly as a result of the formerly exclusive application of chemotherapy to patients with advanced disease and a limited life expectancy. With the identification of increasingly active drugs and drug combinations, however, major anti-tumor effects can be achieved by chemotherapeutic agents.

Although the effect of chemotherapeutic agents is often immediate (like that of antibiotics), objective tumor regression may not be observed clinically for some time. This factor has contributed to the idea that such drugs are ineffective. Time is needed, however, for removal of dead tumor cell debris and for reconstitution of the normal tissue. For example, in a tumor which is metastatic to the liver, the liver scan may require many months to return to normal while all other indicators of the presence of the disease in the liver, including biochemical and serologic abnormalities, will reveal absolute normality.

There are two general principles or concepts of treatment with chemotherapeutic drugs that deserve comment. First, there is an inverse relationship between the expectation of response and the bulk or size of the tumor. The larger the disease burden the less the likelihood of response. This generalization, however, has many exceptions and in patients who have exquisitely sensitive tumors, such as lymphoma, the bulk of disease has relatively little impact on the subsequent response. A second generalization relating to the use of drugs is the correlation of response with both the pathologic and biologic growth rate of the tumor. Pathologic growth rates are measured by the mitotic rate or degree of anaplasia while biologic growth rates are determined by a specific measurement of tumor size over a specific period of time. There is probably a direct correlation between the potential for response to therapy and the growth rate of the tumor, which is measured either pathologically or biologically. It should be emphasized, however, that there are many exceptions to this basic rule and that the rule has never been subjected to scientific investigation.

We will now present some brief case histories of patients who have experienced major anti-tumor effects with chemotherapeutic agents. As these examples reveal, a patient's response to treatment is related to the site of metastasis.

Cutaneous metastases

A 50 year old woman with en cuirasse metastatic breast cancer (Figure 4.8a) was treated with a 5 drug combination

61

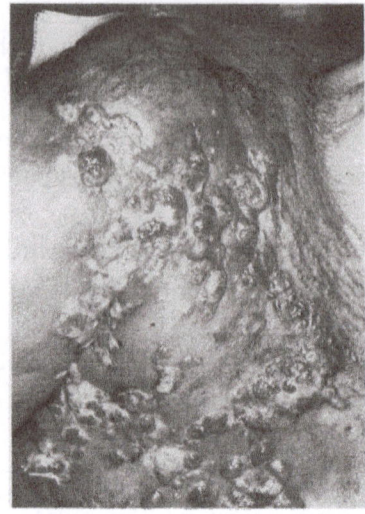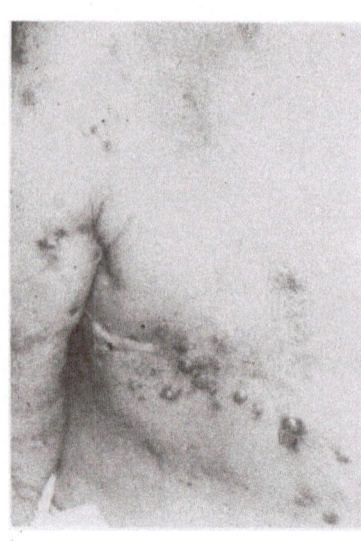

a   b

Figure 4.8
a) Extensive pedunculated lesions, which were entirely neoplastic, secondarily obstructed the skin lymphatics and caused major edema of the arm and chest wall.
b) Chemotherapy induced regression of most of these lesions and promptly relieved the edema and arm pain.

chemotherapy regimen and within 3 weeks demonstrated regression of all but an occasional nodule of tumor (Figure 4.8b). Deep biopsy, however, did demonstrate residual tumor in the subcutaneous tissue.

Pulmonary neoplasia There are primary and metastatic lesions to the lung which may be manifest in a variety of sites including the endobronchial tree, the pleura, or within the pulmonary parenchyma. An example of each follows:

Mesothelioma

A 35 year old woman with biopsy proven mesothelioma was treated with a combination adriamycin and cyclophosphamide and over a 3 month period demonstrated almost complete regression of the tumor (Figure 4.9a).

Endobronchial tumor

A 55 year old patient with oat cell carcinoma was treated with an alkylating agent based combination and within 3 weeks demonstrated radiographic resolution of the right upper lobe collapse (Figure 4.9b).

Pulmonary parenchymal lesion

A 52 year old man with adenocarcinoma to the right and left lung was treated with a combination regimen and within 6 weeks demonstrated major regression of the pulmonary lesions (Figure 4.9c & d).

Hepatic lesions. As indicated previously, hepatic lesions resolve slowly because of the necessity of reconstitution of normal tissue. As a consequence, liver scan defects often persist for a prolonged period of time.

a before   b before   c before

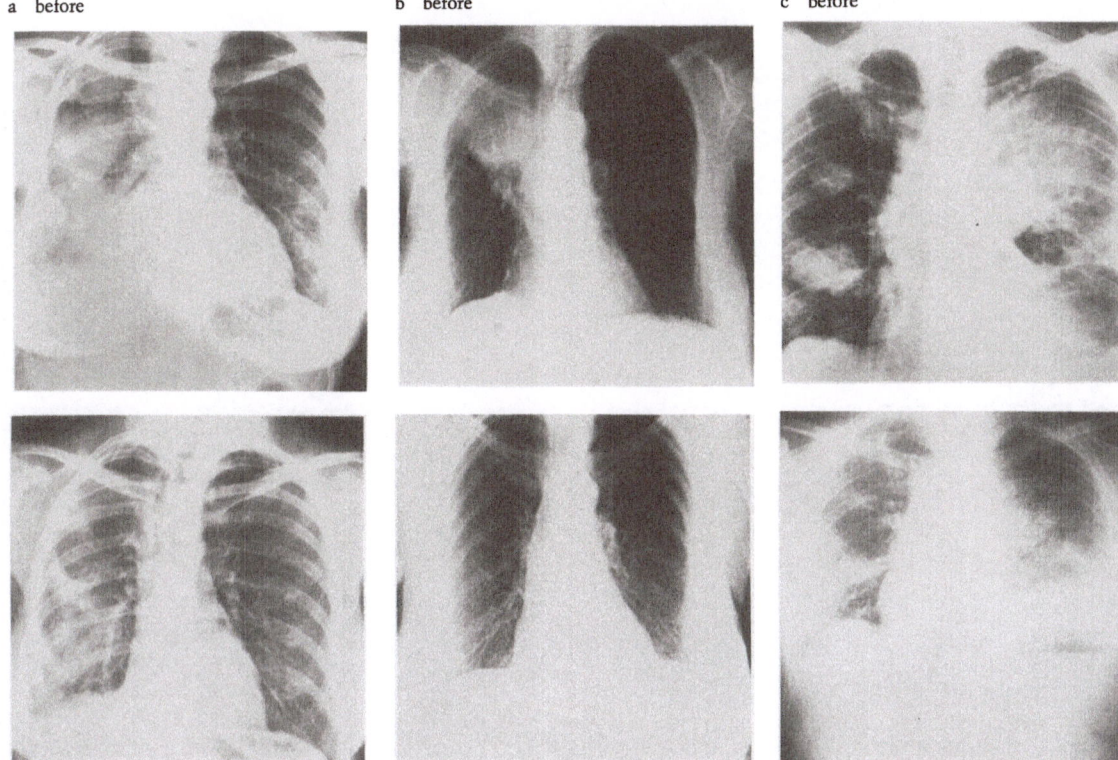

a after   b after   c after

Figure 4.9
a) Large pleural implants which were clinically occult and caused very little symptomatology regressed completely with chemotherapy and only minor residual fluid remains on the *right*.
b) Obstruction to the *right* upper lobe resulted in lobar collapse. However, with chemotherapy and tumor regression, the lung expanded to its normal contour.
c) Adenocarcinoma of the lung, involving multiple sites, regressed extensively because of chemotherapy. A major reduction in the tumor cell burden was achieved.

A 40 year old woman with metastatic breast cancer to the liver was treated with combination chemotherapy and had prompt resolution by physical examination and by tumor marker, but her liver scan improved only over a 12 month period (Figure 4.10).

Brain lesions. Such lesions are difficult to diagnose clinically and quantitatively; the computerized tomogram or the radionuclide scan are suboptimal.

The case of a 40 year old woman with metastatic ovarian cancer to the brain demonstrated a solitary lesion which following radiotherapy resolved completely by brain scan.

Osseous metastasis. Bone lesions are the most difficult metastases to evaluate because the radiographic appearance of the normal bone reaction to the tumor is similar to the healing process in response to tumor cell kill.

An example of a patient who underwent combination chemotherapy for metastatic bone lesions which were primarily lytic in character is illustrated in Figure 4.12. The lesions demonstrate major osteoblastic activity which in association with the healing of other lesions in the skin was interpreted as representing major response to her systemic treatment.

These examples reflect the impact of chemotherapy on a tumor cell population or lesion in a variety of tissue sites.

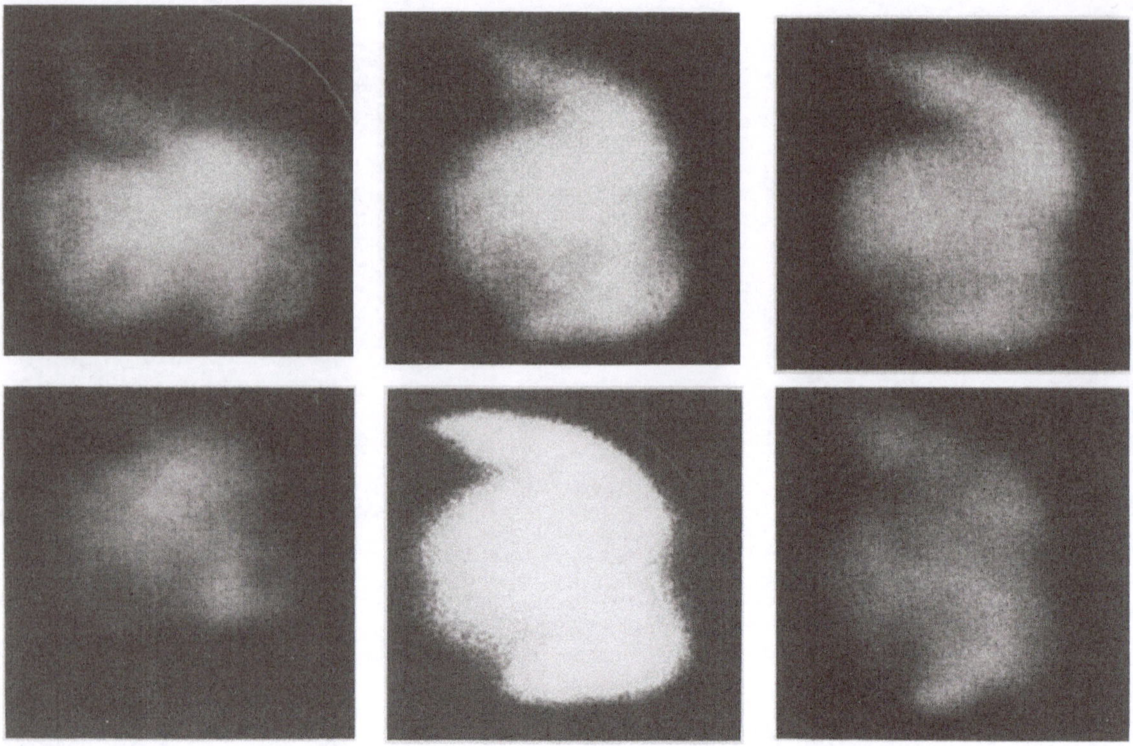

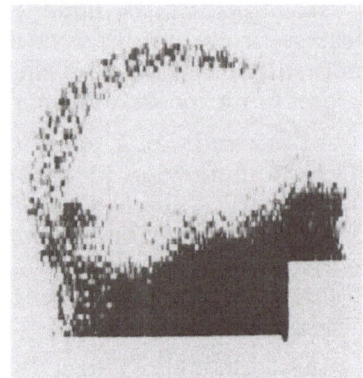

Figure 4.10
Multiple sites of tumor are reflected by the decreased accumulation of the scanning isotope on the liver scan. After chemotherapy, the "holes" filled in slowly as the normal liver tissue was regenerated.

Although the rate and extent of tumor regression varies, any objective diminution of its size means improvement and prolonged survival for the patient.

8.0

Future Prospects

In spite of the multiplicity of drugs that are cytotoxic for tumors and the increasing availability and development of such drugs in the pharmaceutical industry and at the National Cancer Institute, it is unlikely that drug therapy will, in the final analysis, be the complete solution to cancer management.

Figure 4.11
The brain metastasis (indicated by the arrow) receded over a relatively short period of time. Although the lesion eventually recurred, the patient enjoyed a long period of full functioning.

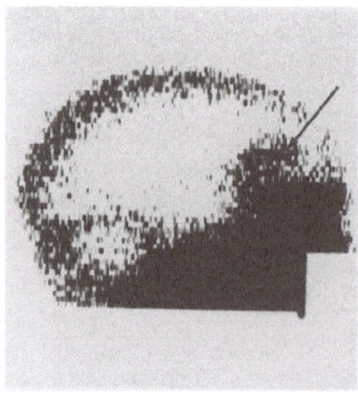

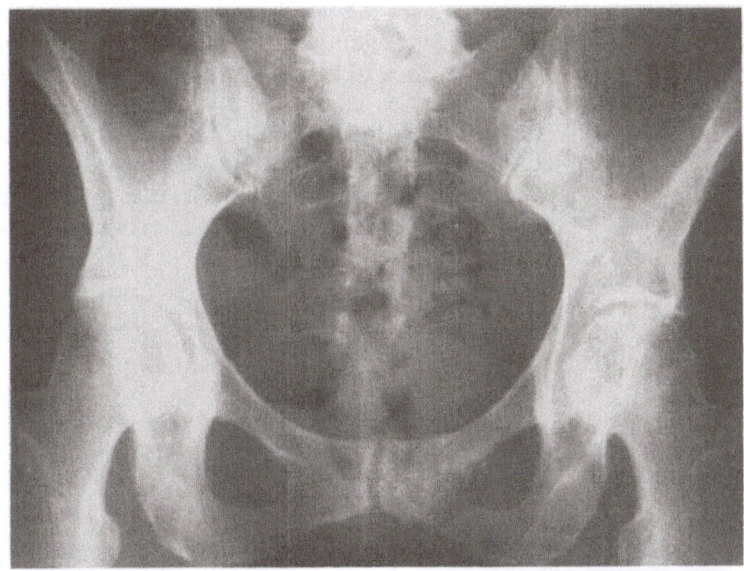

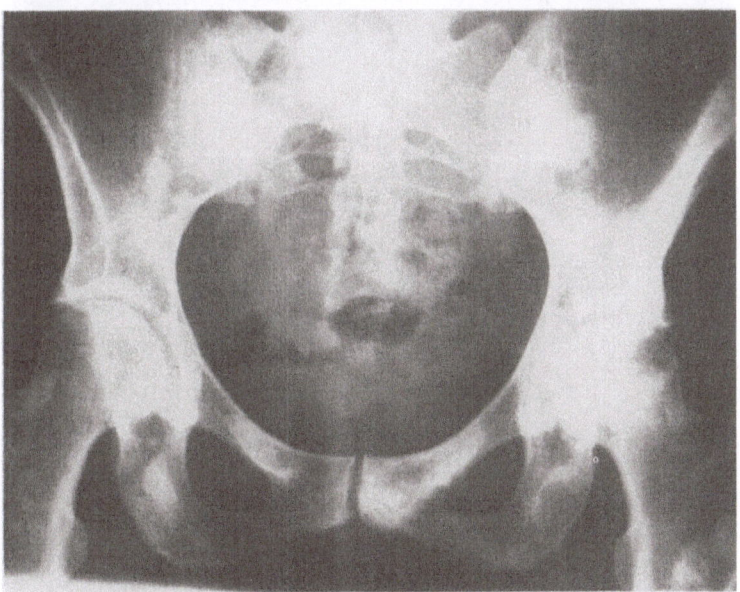

Figure 4.12
The bone metastases of breast cancer were predominantly lytic in this patient as the radiolucent areas in the bone reveal. Following therapy, the radiolucent areas became opacified, representing an osteoblastic healing response.

Nonetheless, the evolution of new drugs provides the potential opportunity for identification of tumor-specific drugs. Other routes of research are developing. In addition to drug combinations (polychemotherapy), drug dose and schedule modification as well as methods of administration may increase tumor cell kill modestly. The critical answer to cancer treatment will be established, however, only if a tumor-specific therapeutic agent can be identified.

The use of chemotherapy in adjuvant circumstances at a time when the host tumor volume burden is minimal may augment the potential effectiveness of drug therapy. Explorations of adjuvant therapy, in breast cancer particularly, have suggested that it may decrease the recurrence rate and prolong survival. Such adjuvant therapy is presently being explored in colon cancer and other forms of gastrointestinal, ovarian and lung cancer, as well as in a multitude of other tumors. Major questions still remain to be answered for adjuvant therapy including the proper duration of therapy, the role and interaction of chemotherapy with radiation therapy and immune therapy, and the long-term potentially adverse effects of such therapy.

References

Brodsky, I., and Kahn, S. B. (eds.), *Cancer Chemotherapy II: The 22nd Hahnemann Symposium.* Grune & Stratton, New York, (1972).

Cancer Chemotherapy: Fundamental Concepts and Recent Advances. Year Book Medical Publishers, Chicago, (1975).

Carter, S.K., Bakowski, M.T., and Hellmann, K., *Chemotherapy of Cancer.* J. Wiley & Sons, New York, (1977).

Greenwald, E.S., (ed.), *Cancer Chemotherapy: Medical Outline Series.* Medical Examination, New York, (1973).

Greenwald, E.S., (ed.), *Supplement to Cancer Chemotherapy: Medical Outline Series.* Medical Examination, New York, (1974).

Livingston, R.B., and Carter, S.K., *Single Agents in Cancer Chemotherapy.* Ifi/Plenum, New York, (1977).

5

Radiation Therapy: General Principles & Techniques

Chapter 5
Radiation Therapy:
General Principles &
Techniques

1.0

Background & Historical Perspective

Radiation therapy has emerged as an independent medical specialty in the United States only in the past 15 years. Radiation therapists are broadly based oncologic physicians with special training in the technical aspects of radiation therapy, radiation physics, and radiation biology. Radiation therapy is a major form of curative treatment, either alone or in combination with chemotherapy and/or surgery, as well as a major palliative tool in the treatment of malignant disease. The use of ionizing radiation dates to Röntgen's discovery of x-rays (1895) and to the discovery of natural radioactivity by the Curies and Becquerel. There were numerous aborted attempts to use this new energy to treat patients. In 1913, Coolidge developed an x-ray tube with heated tungsten filament and target in an effective vacuum producing effective x-rays, and thereby providing the foundation for external radiation therapy. The application of radiation therapy by radium containers and interstitial radium for uterine cancer were developed during this period. In 1922, Regaud, Coutard, and Hautant presented evidence that radiation therapy could be applied to cure advanced laryngeal cancer. Coutard and Baclesse developed a specific modality of protracted fractionation (administering the total dose in smaller daily doses extended over longer periods of time) which is today the basis of radiation treatment schedules. Radiation therapists in France, Scandinavia, and England developed techniques for the treatment of patients with cancer of the cervix and defined the guidelines for local radium application.

With the development of supervoltage radiation machinery, it became possible to deliver a larger dose of radiation to a deep-seated tumor than to the skin, and therefore produce an effective therapeutic ratio. Modern radiotherapy has emerged from this technological advance, and has combined with improved interstitial techniques. A variety of categories of malignant disease, including head and neck cancer, cancer of the female reproductive system, hematologic malignancies, and many other tumors have begun to demonstrate improved cure rates.

Radiation Physics

Technical modalities used in radiation therapy may be separated into three categories. Firstly, external irradiation from sources at a distance from the body is administered by cobalt or higher energy sources such as a linear accelerator or betatron. Secondly, local irradiation is applied by sources in direct contact with a tumor such as the uterus, vagina, and maxillary antrum. Interstitial irradiation refers to the application of removable sources, such as ^{226}Radium, ^{137}Cesium, ^{60}Cobalt, or ^{192}Iridium, inserted directly into the tumor, or non-removable sources such as radon, radioactive gold, or ^{125}I. Thirdly, liquid isotopes, such as ^{32}P and ^{131}I, may be administered systemically or in colloid form into body cavities, such as the pleural or peritoneal cavity. The use of non-removable radioactive sources is possible since they have a relatively short half-life—that is, they lose their radioactivity in a short time.

External irradiation is divided by superficial radiation, orthovoltage, and supervoltage. The energy and penetrating power of ionizing radiation increases as the photon wave length decreases. Differences in physical characteristics of the radiation are of major import. The physical characteristics change over the energy spectrum. The clinically important changes occur with radiation generated in the range between 400 and 800 KV. Above this energy the advantages are reduced absorption in bone, less damage to the skin at portal entry, and better treatment characteristics, such as reduced lateral scatter of radiation into other tissues. Ionizing radiation of sufficient energy to have the characteristics referred to as supervoltage is generally over 500 KV. Ionizing radiation generated at lower energies is termed orthovoltage radiation. The lowest energy radiation is designated as superficial (50–140 KV). Supervoltage radiation has been of great importance in treating tumors deep in the body because skin tolerance no longer limits the dose to be delivered. This method is so effective because the maximum ionization with supervoltage radiation occurs below the level of the epidermis. In addition, since more energetic photons are employed, the secondary electrons they generate travel a great distance in the absorbing material. Therefore, the percentage of radiation of any specific depth compared with the surface dose increases as the energy increases, promoting a therapeutic advantage.

Another method of achieving greater doses delivered to a tumor is to employ multiple fields or to rotate the machine so that normal tissue surrounding the tumor receives less dose while the tumor receives the maximum dose (Figure 5.1).

A selective advantage of the orthovoltage and superficial radiation machine may be achieved in treatment of skin tumors

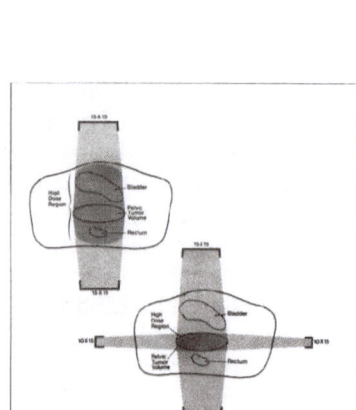

Figure 5.1
Radiation delivery to a deep-seated tumor in the pelvis. The use of 4 fields (b) in contrast to 2 fields (a) results in less dose to normal tissues (bladder and rectum) while maintaining the dose to the tumor.

2.0

in which the depth of penetration will be focused on the skin and preclude deep organ radiation.

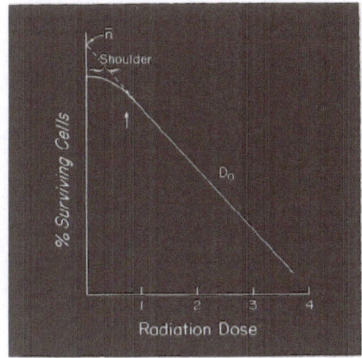

Figure 5.2
Radiation dose-survival curve: Increasing radiation dose results in exponential cell death. D_0 is an indication of the radiosensitivity and is the slope of the straight line portion of the survival curve. The "shoulder" is measured by extrapolating the slope to the ordinate and is referred to as $\bar{n}$ or the extrapolation number.

3.0

Radiation Biology

Contrary to popular belief, radiation does not kill cells by heat but rather does so by a physical chemical change induced by ionization and excitation of molecules in the target tissue. With photon irradiation, secondary charged particles are formed by the transfer of kinetic energy through collision with molecules of absorbing materials; electrons are ejected and ionized molecules result. These chemical products are initially clustered in high concentrations in the path of the charged particles and they interact among themselves or diffusely throughout the absorbing material (tumor). Biologic effects of radiation are proportional to the amount of energy absorbed.

Radiocurability refers to whether a tumor is curable by a given dose of radiation. Radioresponsive refers to the rate of regression but not to whether a tumor is radiocurable. Radiosensitive is a laboratory term that refers to D_0 or the slope of a straight line portion of a radiation survival curve when data is plotted semilogarithmically with surviving fraction plotted on the ordinate and the dose in rad on the abscissa (Figure 5.2). A specific radiation dose kills a constant fraction of cells irradiated. Cells must divide in order to express the radiation lethality effect. The exceptions to this rule are the small lymphocytes and type A spermatogonea which die an intermitotic death. The principle of a fixed or constant fraction of tumor cell kill is illustrated by a cumulative dose survival curve (Figure 5.3). As the dose of radiation increases, the fraction of cells surviving decreases.

The critical radiation damage to the structure of the cell is probably in the DNA, and like chemotherapeutic agents, radiation is cell cycle specific. Thus cells in G_2 and M (the premitotic gap and mitosis) are more sensitive to radiation than cells in other phases of the cycle.

A major concept in the treatment procedure for radiation therapy is the dose and schedule or the fractionation of radiation. A single large dose is as a rule more damaging to normal tissues than doses given in multiple fractions over a long protracted period. Thus, one major rationale for fractionating doses is to allow recovery of the normal tissue.

Another concept of radiation biology which appears to be important in clinical therapy is radiation resistance relative to hypoxia. It has been demonstrated that normal as well as tumor cells are more radioresistant when irradiated under hypoxic conditions. Fractionated irradiation may result in reoxygenation of hypoxic cells, and therefore may be important in clinical radiation therapy.

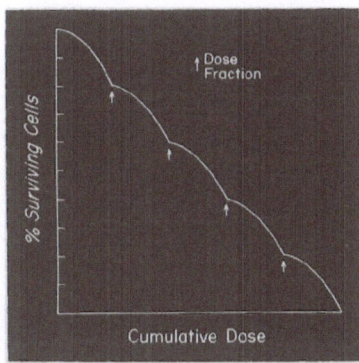

Figure 5.3
Fractionated radiation dose-survival curve: with each dose the shoulder part of the single dose survival curve is reproduced. However, increasing cumulative dose results in effective potential cell death.

4.0

Treatment Procedures & Planning

Treatment planning is essential for optimal radiation therapy delivery. Planning of a course of radiation therapy involves establishing the field of radiation through which the radiation therapy will be delivered. Guidelines to the anatomic localization of the tumor may be necessary. For example, to determine the localization of the tumor in the retroperitoneum one may employ computerized tomography, radionuclide scanning, or ultrasound scanning in association with a contrast radiography. In addition, the surgeon may be requested to place radio-opaque clips in and around a tumor mass in order to delineate the extent of the tumor on radiographs taken during treatment planning.

The use of a simulator is essential and allows the radiation therapist to accurately plan treatment under direct fluoroscopic visualization and to transfer that information to the treatment machine.

Planning is therefore a technique to insure that the tumor receives an optimal dose and that the normal tissues receive as little radiation as possible. By planning one may employ a variety of techniques to avoid adverse effects on normal tissues. For example, lead shields may be developed to protect the lungs when radiation therapy is administered to the mediastinum. Such lead shields are referred to as blocks and are, in fact, employed in a variety of circumstances (Figure 5.4).

The actual treatment delivery is comparable to simply having a standard x-ray taken. Patients enter a protected room within which the x-ray machine's beam is simply focused on the predetermined (planned) portal. After a variable exposure time the treatment is ended. There is no associated discomfort,

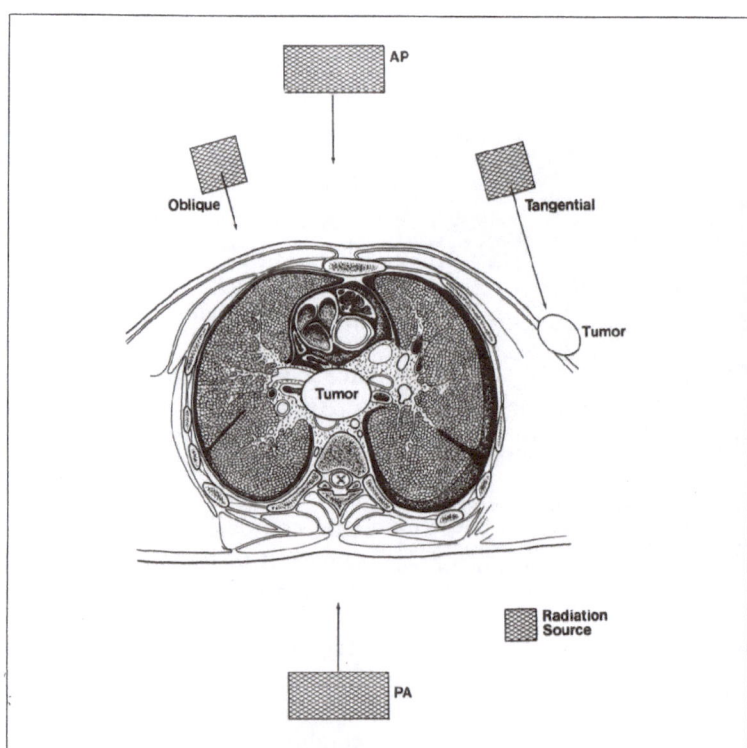

Figure 5.4
Lead shields or blocks protect the normal tissue around the tumor. Two opposed fields (AP and PA) are employed. The lead shields are placed at the source of radiation and protect the lung. Positioning of the radiation source (oblique and tangential) also minimizes normal tissue exposure.

pain, or burning. Treatment is administered basically over a period of 4 to 5 days weekly and depends upon a variety of factors including patient tolerance and tumor characteristics. The rate of administration varies from 150 to 300 rad per day. Most treatment regimens involve 4 to 6 weeks of treatment. In some tumors the treatment regimen may involve an interruption for rest and recovery, thus extending the overall period of treatment by 2 to 4 weeks.

5.0   **Treatment Concepts**

In terms of the therapeutic role for radiation therapy in the management of patients with malignant disease, two broad categories may be established. Radiation given to patients for cure is called radical treatment and with few exceptions, this treatment requires high dose radiation therapy. The second category of patients generally represents patients who are treated for palliation of specific localized symptoms. The dose of radiation in this instance is often less than that one would employ for cure, since production of side effects is undesirable and complete tumor eradication is frequently unnecessary. Increasingly, however, the role for radiation therapy has been expanded to provide primary therapy in the cure of cancer, and to serve as an adjunctive measure to improve local control and maximize curability. The specific role for radiation therapy in a curative, adjunctive, or palliative posture is often related to the stage of the tumor. In the more advanced disease, radiation therapy is primarily a palliative tool as are all therapeutic modalities. With localized or regionalized disease, radiation therapy can often be curative, used alone or adjunctively.

6.0   **Tumor-Specific Application**

6.1   *Oral Cavity, Esophagus.* Radiotherapy alone and in combination with surgery plays a major role in the treatment of oral and oropharyngeal cancer. Tumors are classified using the TNM system of either the American Joint Committee or UICC. T refers to primary size, N to whether regional nodes are involved, and M to distant metastasis. The size of the primary and the presence of nodal and metastatic lesions are taken together to provide a stage that is used as a guide to therapy. External irradiation and implantation, or external irradiation in combination with surgery, is the treatment of choice for carcinoma of the anterior two-thirds of the tongue and the floor of the mouth. Results are approximately the same for external irradiation with an implant or irradiation with surgery; however, the cosmetic results with radical radiation therapy are superior to a large surgical procedure. Approximately 80 to 90% of patients with T_1 lesions can be cured. This figure decreased to 40 to 60% for T_2 and T_3 lesions. Larger lesions require higher doses and the complication rates rise. Carcinoma of the base of the tongue has a much more ominous prognosis. This is because of a luxuriant lymphatic drainage as well as the great difficulty of doing an adequate implant.

73

Surgical procedures are quite mutilating as a total glossectomy is usually required. The overall prognosis of carcinoma of the base of the tongue is approximately 20%, and in general, the treatment of choice is radiotherapy. Carcinoma of the buccal mucosa is successfully treated with radiotherapy either by implantation alone or by external irradiation and implant. It should be noted that bone invasion connotes an ominous prognosis for patients treated with irradiation alone and, in general, if any oral cavity lesion involves bone, irradiation and surgery are required together.

Carcinoma of the palatine arch, retromolar trigone, and tonsil are also successfully managed with external irradiation alone or external irradiation and implantation. The results with preoperative radiation therapy and composite resection (a procedure which removes part of the mandible, oropharynx, and neck) are comparable to radiation alone, but again, the cosmetic results with radiation therapy are superior to combination therapy. Carcinoma of the soft palate and uvula are also well treated with radiation therapy alone.

6.2   *Neck Nodes.* The treatment of metastatic neck nodes continues to be a problem. Previously, radical neck dissection alone was considered adequate treatment; however, recent data suggests that preoperative radiotherapy to a total dose of 4000 to 4500 rad adds to local control. Evidence has emerged that small moveable nodes (N_1) can be treated by radiotherapy and implantation. It has been shown that 4500 to 5000 rad delivered to the neck in N_0 patients prevents the development of metastasis and eliminates the consideration of a prophylactic neck dissection.

6.3   *Salivary Glands.* In general, all head and neck tumors are epidermoid carcinoma. However, malignant salivary gland tumors are divided into malignant mixed tumors, cylindromas, and the mucoepidermoid carcinomas. A characteristic of cylindromas is a proclivity to travel along nerve root sheaths. Although radical surgical procedures were previously performed for salivary gland tumors, the facial nerve can frequently be preserved with excision of gross tumor and radical radiotherapy. It is essential that the base of the skull be covered by radiation protals in all cylindromas.

6.4   *Endolarynx, Hypopharynx, Paranasal Sinuses.* Radiotherapy is the treatment of choice for a carcinoma of the nasopharynx. This is a surgically inaccessible region and successful treatment requires knowledge of the anatomy of this region. Cure rates in early lesions are between 50 and 60% and in advanced lesions between 15 and 20%. Involvement of the base of the skull is frequent. Most tumors that arise in the nasopharynx are squamous cell carcinoma; however, there is a subtype lymphoepithelioma (transitional cell carcinoma) which is more radiocurable than squamous cell carcinoma although there is a slight increase in the incidence of distant metastasis. Occasionally, lymphoma may occur in the nasopharynx and is treated with radiation therapy.

74

Carcinoma of the antrum continues to be a difficult clinical problem. In general, this type of carcinoma is managed with combinations of preoperative radiotherapy and surgery or surgery and postoperative radiotherapy. The anatomic extent of this tumor frequently makes treatment difficult.

Radiotherapy is the treatment of choice in carcinoma of the larynx. The larynx is divided into supraglottic, glottic and infraglottic regions. Early lesions of the true cords have an 80 to 85% cure rate with radiotherapy alone. Approximately 50 to 60% of the radiation failures can be salvaged thus giving a high overall cure rate. Advanced lesions are less curable with radiotherapy; however, many clinicians feel that early T_2 lesions can be managed with radiotherapy and salvage surgery if necessary. Carcinoma of the hypopharynx if in the pyriform fossa is managed with preoperative radiotherapy, laryngectomy or laryngo-pharyngectomy. Carcinoma of the pharyngeal wall is usually managed with radiotherapy alone. The results from these lesions are not satisfactory, however, and the cure rates are relatively low.

6.5   *Carcinoma of the Esophagus.* The diagnosis of carcinoma of the esophagus is rarely established until the tumor is extensive. Rotational or multiple field technique to a total dose of 6000 to 6500 rad in 6½ to 7½ weeks can be delivered. Cure rates are extremely low, approximately 5%. However, significant palliation can be expected. Overall prognosis appears not to be altered whether surgery, radiotherapy, or a combination of treatment is employed. Patients with carcinoma of the cervical esophagus are probably better treated with radiotherapy. Relief of dysphagia or the inability to swallow saliva can provide significant palliation for these patients. Radiation causes changes in salivary function. The patient may complain after his first treatment that painful swelling has developed and that eventually the saliva becomes thick and scanty. If doses are high (over 5000 rad) the salivary function may disappear almost entirely. Therefore, pretreatment dental care is essential, since lack of saliva results in dental caries. All severely carious teeth should be extracted and any infection treated. If root canal treatment is to be done, it should be initiated and completed relatively rapidly since insult to the alveolar bone does not heal well after high dose radiotherapy. Mandibular necrosis must be avoided at all costs. It is essential in the treatment of head and neck cancer that a dentist be involved in a multidisciplinary approach with the radiotherapist and head and neck surgeon. Frequently, fluoride treatments are helpful in preventing radiation induced caries.

6.6   *Genito-urinary (Male).*

Testis. Malignant germ cell tumors are divided into several histologic classifications. These include: 1) seminoma, 2) embryonal carcinoma, pure or with seminoma, 3) teratoma, pure or with seminoma, 4) teratoma with embryonal carcinoma or choriocarcinoma or both, and 5) choriocarcinoma, pure. There is little argument over the treatment of patients with seminoma by inguinal orchiectomy and post radiotherapy to draining lymphatic pathways. Doses vary from 2500 to 3000 rad. Seminoma is one of the most radiocurable diseases in humans. Cure rates of 85 to 90% can be expected in Stages I and II. Appropriate treatment for Stages I and II of teratocarcinoma, embryonal carcinoma, or choriocarcinoma is presently debated. Retroperitoneal node dissection with or without radiotherapy has been frequently employed. Doses range from 4000 to 4500 rad. Recently, with the advent of more effective chemotherapy, patterns of management in this disease may change. Cure rates with Stage I in nonseminomatous disease (that is, confined to the testis with negative lymphangiogram) may be as high as 80% with radiation therapy and surgery. The cure rates drop with advanced stages.

6.7   *Bladder.* Up to 90% of urothelial tumors of the bladder can be designated as transitional cell carcinoma. These tumors are frequently multiple and may be papillary. Staging depends upon the depth of bladder wall invasion. Patients with only superficial or submucosal invasion have a better prognosis than patients with invasion of the deep muscular layers or pelvic organ. Fulguration is generally employed for very superficial tumors and total cystectomy for more advanced tumors. The best results for moderately advanced tumors with deep muscle invasion or penetration of the wall include preoperative radiotherapy and surgery. However, cystectomy is a morbid procedure with ileostomy and impotence as major side effects. Overall cure rates with preoperative radiotherapy and surgery in B_2 and C lesions is 40 to 50% and approximately 20 to 30% with radiotherapy alone. Again, the risks versus gain argument must be employed in the final therapeutic determination.

6.8   *Prostate.* Carcinoma of the prostate is a major cause of death and morbidity in men. The incidence increases with advancing age and is the most frequent cancer in men over 75 years of age. Much discussion has surrounded treatment of this disease since surgical procedures, although effective in obtaining control of early stage disease, have significant side effects (impotence and incontinence). Furthermore, in some men over 65 with well-differentiated lesions, the disease may be biologically indolent, making the management of the localized disease controversial.

Aggressive treatment of disease which is localized to the prostate or the periprostatic tissue with radiotherapy has been highly successful. The 5 year survival rate in patients with disease confined to the prostate is 72% and with extension beyond the prostate, but still localized to the pelvis, 48%. Although some patients become impotent, impotence occurs less often than when radical surgical procedures are employed. Medical and radiation oncologists do not consider hormonal therapy curative. The management of carcinoma of the prostate is undergoing further study by all those concerned with the care of these patients.

6.9   *Soft Tissue Sarcomas.* The role of radiation therapy in the management of patients with sarcoma of the soft tissue has expanded over the past decade. Lesions on the extremities may be managed by local excision and high dose radiation, thereby avoiding amputation. Low grade rarely metastasizing tumors (desmoid, infiltrating, and neurofibroma) not amenable to surgical excision can be treated successfully with high doses of radiation. Pediatric soft tissue sarcoma, such as rhabdomyosarcoma, are now being treated effectively with combinations of radiation therapy, multi-drug therapy, and limited surgery. The management of soft tissue sarcomas in adults with excision and radiation continues to be controversial and only long-term follow-up will give the ultimate answer. Again, meticulous radiotherapeutic technical modifications are necessary and a careful delineation of the extent of the tumor is essential. It is necessary to spare a skin strip during radiation therapy to insure lymphatic drainage from the distal part of the extremity.

6.10   *Breast.* Although radical mastectomy has been standard treatment for carcinoma of the breast, it has been recently challenged by the use of radiotherapy for local disease. Radiation therapy with lumpectomy but without mastectomy is local treatment which offers potential for local control with minimal functional or cosmetic impairment. Here, as in other diseases, technical modifications are of extreme importance to insure homogeneity of the dose throughout the breast. Implantation with interstitial radioactive material is also important. In general, 5000 rad is given with an implant of 1500 to 2000 rad to insure good cosmetic results in Stage I and Stage II. In Stage III lower local control rates are obtained. However, many of these patients are inoperable by surgical criteria. Distant metastasis and survival appear both related to the size of the primary lesion as well as to the status of the axillary lymph nodes. Distant spread cannot be controlled by either mastectomy or local radiation; it requires systemic therapy (see chemotherapy chapter).

If radical mastectomy is performed, the status of post-operative radiotherapy is at present controversial. Local control can be enhanced; whether this will enhance survival, however, is yet to be defined. Several randomized trials have demonstrated both an increase and a decrease in survival. Delineation of groups at high risk for local recurrence after mastectomy will resolve much controversy surrounding post-operative radiotherapy.

6.11   *Central Nervous System.* Most primary intracranial tumors are gliomas and 75% are astrocytic. Glioblastoma multiforme is the most frequent and most biologically aggressive glioma and involves the cerebrum of males from 30 to 50 years of age. In general, the outlook for these patients is quite grim with the overall survival in glioblastoma being less than 5%. Aggressive treatment should not be employed if a good functional recovery cannot be expected. In the well-differentiated astrocytomas, excision and irradiation may provide a significant number of cures.

Ependymomas are tumors that predominantly occur in children and young adults and arise from ependymal cells lining the ventricles of the brain and the central canal of the spine. Surgical removal is rarely complete and such attempts may result in unacceptable operative morbidity and mortality. Many ependymomas respond well to radiation. Long-term survivals may be expected. Doses of 4500 to 5500 rad are used depending on the volume. Cure rates vary from 20 to 60% depending on the grade of the tumor with high grade tumors (poorly differentiated) doing less well.

Medulloblastoma is a highly malignant tumor comprising 25 to 35% of intracranial tumors in children. These tumors arise in the midline of the cerebellum in children, and in the young adult may arise from the cerebellar hemisphere. The medulloblastoma (as does ependymoma) has a tendency to seed the cranial-spinal axis with implant metastasis. Cure rates of 30 to 40% may be obtained with the irradiation of the entire cerebrospinal axis. Whole brain doses of 4000 rad with a cone down dose to the posterior fossa of 5000 to 5500 rad is used. The spinal cord usually receives 2500 to 3000 rad prophylactically. Some authors advocate this form of treatment for ependymomas also. Brain stem glioma is a difficult problem since biopsy may be impossible. Local irradiation should often be preceded by a surgical shunt which may be in itself a major palliative procedure. Cure rates vary from about 20 to 30%.

6.12   *Female Reproductive System.*

Most malignant tumors that arise from the endometrium are adenocarcinoma. Prognosis depends upon the differentiation of the tumor, the invasion of the myometrium, the spread of the tumor to the local pelvic organs, and distant metastasis. Radiation is used in combination with a hysterectomy in Stages I, II, and III. Cure rates in early stages of endometrial cancer are extremely high, ranging from 80 to 90%. Combination therapy has improved the outlook of Stage II and Stage III.

Radiation therapy is the major therapeutic modality for invasive carcinoma of the cervix. Either surgery or radiation therapy may be used in Stage IB; however, radiation therapy is the treatment of choice in all other stages. Staging includes chest x-ray, intravenous pyelogram (IVP), examination under anesthesia, cystoscopy, and proctoscopy. The staging system will be repeated here: Stage 0—in situ. Stage IA—microinvasive carcinoma. Stage IB—carcinoma confined to the cervix. Stage IIA—to upper two-thirds but not to lower one-third of the vagina. Stage IIB—extending beyond the cervix into the parametrium but not reaching the pelvic side wall. Stage IIIA—lower one-third of the vagina. Stage IIIB—tumor reaches pelvic side wall. Stage IVA—invading, involving mucosa of bladder or rectum. Stage IVB—distant metastasis. Both external and intracavitary radiation therapy are employed in the treatment of carcinoma of the cervix. The most commonly employed intracavitary device in the United States is the Fletcher-Suit apparatus. This machine delivers high dose radiation to the cervix and proximal parametrial regions. Varying combinations of intracavitary and external radiation are employed depending upon the stage of the disease. Cure rates in Stage I range from 80 to 90%; in Stage IIA, 75 to 80%; in Stage IIB 60%; in Stage IIIA approximately 50%; and in Stage IIIB 30 to 35%. Only a few patients with Stage IV are cured. The morbidity from radiation therapy is divided into acute effects such as diarrhea and dysuria. However, more serious long-term sequelae, which include chronic damage to the bowel or bladder, may occur. These complications are rare in treatment for early stage disease and are more common for higher doses required for more advanced diseases. Carcinoma of the vagina has a similar staging system and treatment philosophy to that of cancer of the cervix.

The use of radiation therapy in ovarian carcinoma is somewhat debated. When confined to the pelvis but locally advanced, the combination of radiotherapy with surgery shows a clear therapeutic advantage. However, in very early or very late stages of ovarian carcinoma the uses of external radiation are less well defined. Colloidal gold or colloidal phosphate are employed in early stage ovarian cancer since the peritoneal cavity is considered to be at risk for seeding. Several centers have reported an advantage to this treatment; however, it has never been studied in a randomized, prospective fashion.

Pituitary tumors are treated well with radiotherapy. These tumors are divided into acidophilic and chromophobe adenomas and craniopharyngiomas. Irradiation with heavy particles (alpha particles, protons) has been advocated for various pituitary tumors. However, well planned conventional supervoltage radiotherapy is the most applicable form of treatment. Approximately 80% of both chromophobe and acidophilic adenomas can be controlled with 4000 to 5000 rad. Fifty to 80% of the craniopharyngiomas can also be controlled. This tumor, however, requires higher doses, from 5000 to 5500 rad.

6.13 *Lung.* Carcinoma of the lung kills more males than any other tumor. Various tissue types include epidermoid, undifferentiated, oat cell, and adenocarcinoma. In general, resection is the treatment of choice for adenocarcinoma and epidermoid carcinoma. Frequently, however, resection is not possible. Local radiotherapy may be attempted even in unresectable cases with an expected cure rate from 5 to 10%. Preoperative radiotherapy may be employed in superior sulcus tumors or tumors which remain localized for long periods of time, and a 20 to 30% cure rate may be expected in this group. Oat cell carcinoma metastasizes early and widely. Combined chemotherapy and radiotherapy are now undergoing clinical trials in this disease. Even in the presence of distant metastasis, radiotherapy

79

may be helpful for relief of bronchial obstruction, superior vena caval syndrome, cough, local pain, as well as local symptoms caused by distant metastasis. It is important that superior vena cava syndrome be treated properly with no manipulative procedures in the obstructed region.

6.14   *Lymphoma.*

Hodgkin's Disease is a morphologically diverse malignant tumor of the lymphoid tissue composed of reticulum cell and lymphocyte derivatives. Pleomorphism and cellular diversity have led to a histologic subclassification. This system has been developed through the years and has been evolved to correlate with clinical behavior and prognosis. The subtypes are: (1) lymphocyte predominant, (2) nodular sclerosis, (3) mixed cellularity, and (4) lymphocyte depleted. Lymphocyte predominant has the best prognosis and lymphocyte depleted the worst, with nodular sclerosis and mixed cellularity intermediate in their outlook. The single most important prognostic indication is stage. Stage I is involvement of a single lymph node region or a single lymphatic site. Stage II involves two or more lymph node bearing regions of the same side of the diaphragm. Stage III involves lymph node regions on both sides of the diaphragm which may be accompanied by involvement of the spleen or by solitary involvement of an extralymphatic organ. Stage IV is multiple or disseminated foci of one or more external organs or tissues with or without lymph node involvement. All stages are subclassified as A or B to indicate absence or presence of systemic symptoms. The B symptoms are night sweats, fever or weight loss of greater than 10% of body weight. Advances in radiotherapy and chemotherapy have dramatically altered the prognosis of Hodgkin's disease, once a fatal illness. Radiotherapy, applied in early stages of the disease, cures between 80 and 90% of all patients. The treatment is referred to as "mantle" radiation and includes lymphatic areas in the neck, mediastinal, axillary, and supraclavicular regions. The para-aortic fields include the lymph nodes in the upper abdominal region. Total nodal radiation includes not only mantle and para-aortic nodes but pelvis irradiation as well. MOPP (Mustargen, Oncovin, procarbazine, prednisone) chemotherapy is now being combined with radiation therapy to further improve the outlook of patients with advanced disease (see Chapter 4).

Non-Hodgkin's Lymphoma. Previously referred to as reticulum cell sarcoma, lymphosarcoma, and giant follicular lymphoma, the non-Hodgkin's lymphomas have now been histologically subclassified as to: 1) whether they are lymphocytic and histiocytic, 2) the differentiation of the cells, and 3) whether the architecture of the lymph node involved is nodular or diffuse. Distinct from Hodgkin's disease which appears to have a relatively orderly progression, non-Hodgkin's lymphoma presents itself as advanced disease in approximately 80 to 90% of all cases. However, a modified staging of Hodgkin's disease is used. Localized lymphomas are curable in 50 to 60% of all cases with radiotherapy. Certain presentations such as Waldeyer's ring, GI tract, and primary bone lymphomas carry a better prognosis since these tend to be localized. These are treated with aggressive local radiotherapy and are curable in 50 to 60% of all cases. Whole body irradiation may be employed in nodular lymphocytic lymphomas and chronic lymphocytic leukemia. This is done as a palliative measure, and is as efficacious as chemotherapy and not as toxic.

6.15

Skin. Three major pathologic types occur in skin cancer. These are melanoma, a malignancy of the melanocytes, basal cell carcinoma, and squamous cell carcinoma. Although, radiation therapy is used for melanoma in European centers the treatment of choice in the United States is surgical excision. Any skin carcinoma on the extremity or in a non-cosmetic site, if not deeply invasive, may be managed by simple excision. However, when both basal cell carcinoma and squamous carcinoma occur on the face or in other surgically inaccessible sites, radiation therapy is employed. Techniques, as in other areas, are very important as is selection of the proper energy machine (since the depth must be adequately estimated). When proper fractionation is employed (increase in fractionation generally gives better cosmetic results), excellent cosmesis may be expected. Radiation has special advantage for tumors of the skin, nose, eyelids, and ear. Care must be taken to avoid trauma to irradiated skin as necrosis may occur.

7.0

Radiation Therapy Complications

Complications of radiation therapy are related to the dose of treatment, the time of administration of treatment, and the volume of treatment. Acute secondary complications develop as a consequence of radiation damage to the proliferating cell renewal systems, and are therefore most noticeable in tissues with proliferating cell renewal systems. These tissues include the mucosal lining of the oral cavity and pharynx, the lining of the gastrointestinal tract, the bladder mucosa, skin and its appendages, and the bone marrow. Each of these tissues will have a tolerance dose; doses exceeding the tolerance will result in specific signs and symptoms. The tolerance dose is affected by the volume of the tissue treated and the rate of dose administration. There also seems to be individual patient variation which may be related to the patient's overall condition. A specific example of an acute complication is the mucositis of oral irradiation that occurs because of the inability of the basal layers to replace the desquamating superficial layers. The patient will notice pain in the mouth and will have difficulty swallowing. These acute effects are generally reversible and can be managed by altering the dose fractionation. A prolonged schedule of treatment or intermittent treatment may be employed to allow normal tissue recovery.

In contrast to the acute complications of radiation, the chronic effects are most often observed only after a protracted period following radiation and may occur many years following irradiation. The mechanism of chronic radiation complication is related not to interruption of normal cell growth but rather to the secondary effect of radiation on the vascular supply to the tissue. Small arteries become narrowed or occluded resulting in a decreased blood supply to the specific organ supplied. The result can be seen in any organ or tissue, and as with acute effects is related to dose, volume, and rate of administration. Each organ has a unique tolerance to these chronic effects and dose-time limits are available for most organs (Table 5.1).

Table 5.1
Organ Dose Tolerance for Normal Tissue Radiation Effects

Whole Organ	Injury	Dose in Rad for 5% Complication Incidence
Brain	Necrosis	6000
Spinal Cord	Necrosis (Transection)	4500
Peripheral Nerve	Neuritis	6000
Lung	Pneumonitis	1500
Heart	Pericarditis	4500
Kidney	Nephritis	200
Liver	Hepatitis	2500
Gastrointestinal Tract	Perforation, Ulcer Hemorrhage	4500
Bone Marrow	Aplasia	250 (Whole Body)
Salivary Gland	Xerostomia	5000
Muscle (Adult)	Fibrosis	6000
Bone (Adult)	Necrosis	6000
Lens	Cataracts	500

Radiation therapy complications are very important and are frequently the most critical limitation in treatment. Exceeding the tolerance doses to vital organs like the heart, lung, liver, and kidney can potentially result in dire consequences. In the case of non-vital organs, tolerance doses are often exceeded and complications are accepted if the likelihood of cure is significant. Even though tolerance dose is exceeded not all patients will develop problems, since the likelihood of compli-

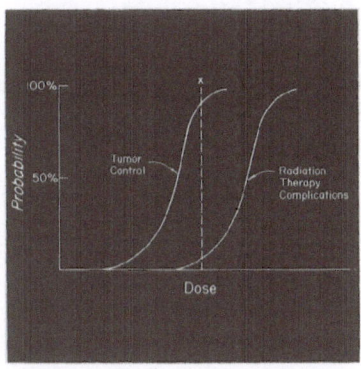

Figure 5.5
Tumor control and complications curves related to dose of radiation. A tumor treated to dose at X is a treatment program with high probability of tumor control and acceptable probability of complications.

cations is a stochastic event. In practice, the probability of complications is balanced against the probability of tumor control (Figure 5.5). This balance allows the radiation therapist to determine the dose of irradiation that will likely control the tumor and the risk of normal tissue reaction produced by this dose.

The fact that radiation can induce tumors is well known. Less well appreciated, however, is the fact that the radiation associated with carcinogenesis has invariably been low dose irradiation or chronic exposure. Therefore, the use of therapeutic radiation employing the supervoltage machines of today is generally not associated with cancer production or induction and should not be a weighty consideration in evaluating the

Table 5.2
Radiation Therapy Complications: Relative Incidence and Reversibility

Complication	Comment
Secondary Malignancy	Associated with long, low dose & low energy exposure; rarely, if ever, associated with therapeutic radiation (see text).
Local Organ Effects	Dose related tissue tolerance.
Head	Brain necrosis >6000R
Spinal Cord	Transection >6000R
Peripheral Nerve	Palsy >6000R
Lung	Pneumonitis & Fibrosis >1500R
Heart	Cardiomyopathy >4500R
Kidney	Nephritis >3500R
Liver	Hepatitis >2500R
Gastrointestinal Tract	Esophacitis, Enteritis, Colitis, Salivary Glands see text
Bone	Osteonecrosis see text
Genito-urinary	Impotence, >5000R Sterility, Mutagenicity see text
Systemic Effects	"Radiation Sickness"

Acute Adverse Effects
  Perforation
  Obstruction
  Edema
	See text.

Long Term Adverse Effects	See text.

need for therapy. There is some data that indicates that chemotherapy employed in conjunction with radiation therapy may increase the risk for the development of subsequent leukemia or other types of neoplasia. Invariably, the neoplasia secondary to radiation evolves at a substantial interval following exposure. The secondary malignancy is often a resistant tumor and difficult to control with today's therapeutic approaches.

References

General

Buschke, F., and Parker, R. G., *Radiation Therapy in Cancer Management.* Grune and Stratton, New York, (1972).

Fletcher, G. H., *Textbook of Radiotherapy.* Lea and Febiger, Philadelphia, (1973).

Hall, E. G., *Radiology for the Radiologist.* Harper and Row, Hagerstown, Maryland, (1973).

Moss, W. T., Brand, W., and Battifor, A. H., *Radiation Oncology.* C.V. Mosby, St. Louis, (1973).

Rubin, P., and Casarett, G. W., *Frontiers of Radiation Therapy and Oncology.* In J.M. Vaeth (ed.), vol. 6, P.I. University Park Press, Baltimore, (1972).

Head and Neck

Berger, D. S., Fletcher, G. H., Lindberg, R. D. and Jesse, R. H., *Am. J. Roentgen.* **111**:66, (1971).

Chen, K. U., and Fletcher, G. H., *Radiol.* **99**:165, (1971).

Fletcher, G. H., Jesse, R. H., Lindberg, R. D. and Koon, C. S., *Am. J. Roentgen.* **108**:19, (1970).

Fletcher, G. H. and Lindberg, R. D., *Am. J. Roentgen.* **96**:574, (1966).

Montana, G. S., Hellman, S., von Essen, C. F. and Kligerman, M. M., *Cancer* **23**:1284, (1969).

Wang, C. C., Schultz, M. D. and Miller, D., *Am. J. Surg.* **124**:551, (1972).

Wang, C. C. and Schultz, M. D., *Radiol.* **80**:693, (1963).

Breast

Levene, M. D., Harris, J. R., and Hellman, S., *Cancer* **39**:2840, (1977).

Weber, E., and Hellman, S., *JAMA* **234**:608, (1975).

Weichselbaum, R. R., Marck, A., and Hellman, S., *Cancer* **37**:2682, (1975).

85

Lung

Guttmann, R. J., *Am. J. Roentgen.* **93**:99, (1965).

Hellman, S., Kligerman, M. N., von Essen, C. F., and Scibetta, A., *Radiology* **82**:1055, (1964).

Mallams, J. T., Paulson, D. L., Collier, R. E., and Shaw, R. R., *Radiology* **82**:1050, (1964).

Skin

von Essen, C. F., *Chicago Year Book Medical Publishers,* 285, (1964).

von Essen, C. F., *Act. Radiologic Ther.* **8**:311, (1969).

von Essen, C. F., *Brit. J. Radiol.* **42**:474, (1969).

Soft Tissue Sarcoma

Haynes, R. M., *Cancer* **35**:921–924, (1975).

Hill, D. R., Newman, H., and Phillips, T. L., *Am. J. Roentgenol.* **117**:84, (1973).

Suit, H. G., Russel, W. O., and Martin, R. G. *Cancer* **35**:1478, (1975).

Weichselbaum, R. R., Cassady, J. R., Jaffe, N., and Filler, R., *Int. J. Radiation Oncol. Biol. Physics.* **267**:72, (1977).

Esophagus

Millburn, L., Hendrickson, F. R., and Faber, P., *Am. J. Roentgen.* **103**:291–299, (1968).

Pearson, J. G., *Brit. J. Surg.,* **58**:794, (1971).

Bladder

Caldwell, W. L., Bagshaw, M. A., and Kaplan, H. S., *J. Urol.* **97**:294, 1967.

Miller, L. S. In: *Oncology* (1970).

Proceedings of the 10th International Cancer Congress, Yearbook Medical, Chicago, (1971).

Whitmore, W. F., Grabstadt, H., MacKenzie, A. R., Oswaria, H. J., and Phillips, R., *Am. J. Roentgen.* **102**:570, (1968).

Prostate

Ray, G. R., Cassady, J. R., and Bagshaw, M. A., *Radiology* **106**:407, (1973).

Testis

Castro, J. R., and Gonsales, M., *Am. J. Roentgen.* **2**:355, (1971).

Castro, J. R., *Cancer* **24**:87, (1969).

Smithers, D. W., and Wallace, E. N. K., *Brit. J. Urol.* **34**:422, (1962).

Female Reproductive—Cervix

Fletcher, G. H., and Rutledge, F. N., In: Thomas J. Dealy (ed.), *Modern Radiotherapy,* p. 11, Butterworth, London, (1971).

Fletcher, G. H., *Am. J. Roentgen.* **225**:242, (1971).

Suit, H. D., Moore, E. B., Fletcher, G. H. and Worsnop, R., *Radiology* **81**:126, (1963).

Endometrium

Delclos, L., Fletcher, G. H., Gutierre, A. G. and Rutledge, R. N., *Am. J. Roentgen.* **105**:603, (1969).

Gusberg, S. B., and Yannopoulos, D., *Am. J. Obstet. Gynec.* **88**:157, (1964).

Ovary

Delclos, L., and Smith, J. P. In: Fletcher, G. H., *Textbook of Radiotherapy*, Lea & Febiger, Philadelphia, (1973).

The Cancer Committee of the International Federation of Gynecology and Obstetrics, *Acta Obstet. Gynec. Scan.*, **15**:1, (1971).

Hodgkin's Disease

Kaplan, H. S., *Hodgkin's Disease.* Harvard Univer., Camb. Ma., (1972).

Non-Hodgkin's Lymphoma

Million, R. In: Fletcher, G. H., *Textbook of Radiotherapy*, Lea & Febiger, Philadelphia, (1973).

6 Immune Therapy: Principles & Techniques

Chapter 6
Immune Therapy: Principles & Techniques

1.0

Introduction & Background

The immunologic approach to the treatment of malignant diseases has been discussed in medical literature and developed on the experimental level for more than a century. During the past decade a variety of immunologic modalities have been used in studies in human cancer patients to control the disease. Immunotherapy is conceived of as a possible coequal or adjunct to chemotherapy and radiotherapy in the treatment of certain malignancies. Although the role of immunotherapy has not yet been clearly delineated, studies in animals and patients serve as a foundation on which to build future treatment programs.

In this chapter we hope to offer a perspective through which to understand both the work that precedes us and the work which will come in the future. It should be kept in mind that the science of immunology in the broad sense as well as the discipline of tumor immunology in particular are advancing rapidly. Many insights into basic immune response and mechanisms of immunologic injury are being gained at an enormous rate. As these insights are applied to the clinic, the immunologic approaches to the treatment of malignant disease should continue to grow in relevance and effectiveness.

Much of the early work in tumor immunology dealt with attempts to "vaccinate" animals against transplanted tumors in much the same way animals were being vaccinated against infectious diseases. These experiments usually involved the vaccination of an experimental animal with a tumor derived from a non-inbred animal. Frequently, the recipient animal rejected the tumor more rapidly if previously "vaccinated" with it. This result was interpreted to mean that a tumor-specific immune response had been generated.

It was not until transplantation immunology developed that the more likely explanation of these experimental results was clarified. Tumors, like other tissues, contain histocompatibility antigens. Non-inbred animals, i.e. *allogeneic* animals, are virtually always non-identical with respect to histocompatibility antigens. Thus, tumors may be rejected in the same way that kidneys, bone marrow, and other normal tissues are rejected, i.e. because of non-identity within the major histocompatibility complex and not because of tumor related antigens. The exploration of tumor immunology awaited the definition of the major histocompatibility antigen systems in animals and humans and the development of inbred colonies of experimental animals which after many generations of inbreeding were identical for all histocompatibility antigens and are termed *syngeneic*.

Modern tumor immunology essentially began in 1957 with the definitive demonstration of tumor-specific transplantation antigens (TSTA) in mice by Prehn and Main. These workers immunized mice against a chemically induced tumor from a syngeneic animal. They demonstrated that following simultaneous engraftment of a tumor and a skin graft from the same syngeneic animal, the tumor would undergo immune rejection while the skin graft would not. From these results could be deduced the presence of a TSTA which was not represented on the skin graft and which allowed for specific recognition and elimination of the tumor.

Further studies in animals suggested that the element of the immune response which appeared to be most important in tumor elimination was the "cellular immune response" as opposed to the "humoral immune response," i.e. antibody formation. This conclusion was reached on the basis of the similarity between "rejection" of experimental tumors and rejection of transplanted tumors both histologically and kinetically.

The demonstration of TSTA coupled with the expanded understanding of the immune system led Lewis Thomas to postulate and subsequently Sir MacFarlane Burnet to elaborate on the theory of "immunologic surveillance." In brief, this theory states that all cells which undergo malignant transformation acquire new antigens, i.e. TSTA. Furthermore, it is a fundamental function of the immune system to recognize these neoplastic cells as antigenically dissimilar from normal "self" and to initiate their destruction. Thus, the immunologic system which operates so successfully to thwart attempts at transplantation is seen as an adaptive system whose primary purpose is to defend the body against the development of cancer.

Two major corollaries of the immunologic surveillance theory have particular importance to the cancer therapist. The first corollary states that the development and spread of cancer represents a failure of the immune system. Thus, in some way the cancer patient is presumed to be immunodeficient and if the appropriate means of measuring the immune response were available, cancer patients, and even patients with an enhanced predilection for developing cancer, could be detected. The immunologic evaluation of cancer patients has therefore become a major concern of cancer research.

The second corollary states that if the immune system can

be appropriately manipulated, then perhaps a variety of immunologic mechanisms might be used to therapeutic advantage in the treatment of cancer. This theory was probably the major impetus to the development of "immunotherapy."

The immunologic surveillance theory initiated an era of enormous experimental and clinical effort in tumor immunology. Interestingly, neither the theory nor its corollaries have ever been definitely proven. Nonetheless, the data which have accumulated in the wake of this theory form a solid and rapidly expanding foundation for exploration of immunological attack on tumors.

Tumor Immunology

2.0

The immune system is basically a dual system. One part of the system is under the general control of the thymus gland. During very early development, the thymus influences the differentiation of a proportion of small lymphocytes which are thought to originate in the bone marrow. The precise mechanism of this influence is not known. Some investigators believe that a product made by the thymus, thymosin, is instrumental in this process. Whatever the mechanism, these cells differentiate into T cells (thymus dependent cells) and are instrumental in such functions as graft rejection and delayed hypersensitivity reactions. It is presently believed that the thymus dependent system is particularly important in tumor rejection and thus the major focus of tumor immunology.

The second part of the immune system is responsible for the synthesis and release of specific antibodies. This function is mediated by cells which are also thought to originate in the bone marrow. The Bursa of Fabricius, a distinct organ found attached to the GI tract in chickens, is responsible for influencing the differentiation of these cells in chickens. They differentiate first into small lymphocytes with immunoglobulin on their surface termed B cells, Bursa-dependent cells. Finally, these cells differentiate into plasma cells which produce circulating antibodies. The Bursa of Fabricius has no clear analogue in humans, but it is thought that the same influence on the differentiation of small lymphocytes is exerted by a variety of lymphoid tissues, especially those which are closely related to the GI tract.

T cells and B cells cannot be distinguished on morphologic grounds since they both appear to be normal lymphocytes. They can, however, be distinguished by relatively simple laboratory procedures. B cells can be detected by making use of the fact that these cells have immunoglobulin on their surface. By reacting peripheral lymphocytes with fluorescinated anti-immunoglobulin, the B cells fluoresce, whereas T cells do not. On the other hand, only T cells have receptors for sheep erythrocytes on their surfaces. They may be detected by incubation with sheep erythrocytes and the formation of "rosettes," i.e. a T lymphocyte surrounded by many sheep erythrocytes where the peripheral erythrocytes are likened to the petals of a flower.

B cells and T cells do not account for all of the mononuclear cells present in the peripheral blood nor all of the mononuclear cells thought to be of importance in immune response. There is a population of small lymphocytes which have been called "null cells" by some investigators. This population of morphologically similar cells with neither surface immunoglobulin nor receptors for sheep erythrocytes contains a diverse group of functionally heterogeneous cells which are only beginning to be understood. Within the null cell population reside human hematopoietic stem cells, pre-B cells, and probably the cells responsible for killing in the antibody dependent cellular cytotoxicity assay (to be discussed later).

A fourth mononuclear cell which can be morphologically distinguished from lymphocytes is the monocyte. The role of monocytes (which become tissue macrophages when they leave the circulation) in the immune process is not entirely clear. There is evidence to suggest that monocytes (or macrophages) are necessary for processing antigens which initiate the immune response. In addition, they are phagocytic cells and have been shown to participate in immune elimination of antibody-coated cells. Much work is in progress to explore further functions of monocytes in immune response.

3.0 **Tumor Antigens**

Tumor related antigens have been discovered in both animals and humans. Their presence forms the essential basis for immunotherapy and their continued investigation is imperative if immunological approaches to cancer therapy are to improve. Whether or not tumor-specific transplantation antigens (antigens present only on malignant cells of a particular type and on no normal cells) have been definitively detected in humans remains a controversy. Nonetheless, most workers feel they are present even if the technology is not yet advanced sufficiently to reliably detect them. For purposes of discussion, tumor related antigens will be discussed under three major headings: viral related antigens, oncofetal antigens, and idiotypic antigens.

Viral Related Antigens

Oncogenic viruses are responsible for two types of new antigens on the surface of infected cells. One antigen is present on the virus itself. The second type, called the T antigens, are present on tumor cells which are not necessarily producing virus particles and cannot be detected on the virus itself. Both of these viral related antigens may be important in immune rejection in animals. However, their role in human malignancy remains undefined because of the uncertainty of the role of oncogenic viruses in human cancer.

Oncofetal Antigens

Oncofetal is the name applied to antigens associated with both tumors and embryonic tissues from which these or related tumors are derived. Carcinoembryonic antigen (CEA) is probably the best known and the most thoroughly studied in humans. This is a glycoprotein molecule which originally was detected in patients with colon cancer. An antisera was prepared against an extract of human colon cancer cells in rabbits. After a series of absorptions, the antisera detected an antigen associated with most human colon cancers and with embryonic colon tissues.

More than a decade of investigation has shown that elevated CEA levels in serum are present in a wide variety of malignancies, a variety of non-malignant inflammatory conditions, and in apparently well people who smoke cigarettes. Thus, clearly CEA is not a TSTA available for specific attack by immunotherapy. However, monitoring CEA remains useful as a tumor marker in assessing the success or failure of a particular therapeutic intervention in certain cancer patients.

Alphafetoprotein (AFP) is a second oncofetal antigen originally associated only with hepatoma, teratocarcinoma, and fetal tissue. Further investigation with more sensitive detection techniques has revealed elevated levels of AFP in a variety of other tumors, such as gastric, colon, pancreatic, and prostatic. More importantly, as with CEA, AFP has been found in low levels in normal individuals. Thus, this antigen is neither tumor-specific nor does it offer any opportunity for tumor-specific attack. As with CEA, AFP has been useful as a tumor marker.

Other oncofetal antigens have been described and, as with CEA and AFP, they are not tumor-specific. Their importance relates to the biologic significance of their association with a particular array of tissues, malignant and benign, and to their adjunct utility in the diagnosis and prognosis of malignancy. They do not offer a target for immunotherapy.

Idiotypic Antigens

Tumors induced by chemicals in animals, such as methylcholanthrene (MCA), develop surface antigens which are termed idiotypic antigens. These antigens are generally identical within a specific tumor but usually differ from one MCA induced tumor to the next. These antigens are tumor-specific and very potent transplantation antigens. Thus, tumors with these antigens are strongly immunogenic and are quite vulnerable to immunologic attack.

From this discussion of what is known of tumor antigens, it is clear that viral related antigens and idiotypic antigens are more likely to be the target for successful immunologic intervention in the control of malignancy. It should be kept in mind, however, that it is likely not all tumors possess strong or even any tumor-specific antigens and thus not all tumors will be vulnerable for immunologic intervention.

4.0   **Mechanisms of Immunologic Destruction of Tumors**
An enormous amount of work is presently being expended in the investigation of tumor destruction by basically immunologic mechanisms. A brief description of the three best docu-

mented *in vitro* mechanisms follows. It should be understood that the list is not meant to be necessarily inclusive nor can the *in vivo* relevance of each mechanism be readily inferred.

Antibody and Complement

Tumor cells, much like normal erythrocytes or normal lymphocytes, can be destroyed following incubation with antibody and complement. The cells are destroyed by the activation of the classical complement sequence. Cell death is generally measured by the release of ^{51}Cr into the medium.

T cell Mediated Cytotoxicity

T cell mediated cytotoxicity is the term used to describe the cellular destruction that results from the interaction between specifically sensitized T cells and their target tissues. The reaction is totally independent of antibody and complement and can be shown to be entirely T cell mediated. The antigens which serve as both sensitizers and targets can exist on either normal or tumor cells.

Antibody dependent cellular cytotoxicity (ADCC)

This mechanism of cell mediated destruction has been known by many different names. It combines the specificity of antibodies directed against cell surface antigens with the cytotoxicity potential engendered by a sub-population of small lymphocytes deemed K cells by some investigators. The antibodies seek out the target tissue and the small lymphocytes recognize the Fc portion of the antibody attached to the target tissue and destroy the tissue to which the antibodies are attached. The effector cell is probably a non-T cell. This process is an extremely efficient means of cellular destruction.

The existence of these three mechanisms of tumor destruction allows one to easily conceive of immunologic approaches to tumor eradication. The immunologic surveillance theory need not be entirely true for the immunologic assault on malignancy to have clinical utility.

5.0

Immunologic Evaluation of Cancer Patients

As mentioned above, the immunologic surveillance theory implies that there may be a defect (or defects) in the immune system of patients with cancer which predisposes them to the development of malignancy. Although patients with a variety of congenital immunodeficiency diseases have an increased incidence of malignancy, no single defect which correlates directly with the development of cancer has yet been discovered. The search therefore continues.

Immunologic evaluation of cancer patients has taken a number of approaches. Thorough study of the end products of the humoral immune system, i.e. immunoglobulin levels and antibody synthesis, has not resulted in any increased understanding of the relationship between malignancy and cancer with the exception of malignancies involving B cells or plasma cells. This type of immunological evaluation will not be discussed in detail. The three modes of evaluation discussed below represent the most popular means of immunologic evaluation. These are recall skin testing with antigens to which the majority of patients have been previously exposed, Dinitrochlorobenzene (DNCB) sensitization and subsequent DNCB challenge, and the evaluation of peripheral blood mononuclear cells.

Recall Skin Testing

Recall skin testing involves the introduction into the skin of an antigenic solution. If the patient has an intact immune response (including "immunologic memory") and has been sensitized previously by the antigen in the solution, then a reaction characterized clinically by induration and histologically by the infiltration of the skin with a variety of mononuclear cells ensues. This is a delayed reaction requiring 24 to 48 hours or even longer to reach peak intensity.

Most investigators use a battery of several recall antigens so that a patient will be challenged with at least one antigen to which he is sensitive. We have found that the combination of mumps antigen and Streptokinase-Streptodornase (SKSD) offers a combination that fulfills this criterion. In general, we also use intermediate strength PPD because many of the patients we are assessing may receive BCG or a product derived from it during their therapeutic course.

Skin tests are a semi-quantitative measure of the "recall" limb of the delayed hypersensitivity response in that the dimensions of induration are in general related to the degree of sensitization of the patient. However, the delayed hypersensitivity response involves much more than immunologic recall function. The intensity of the response can be modulated by a variety of circumstances including the type and intensity of the therapy, the type of malignancy, the coexistence of other illnesses affecting the immune response, and the overall clinical status of the patient. Thus, the serial measurements of skin tests must be carefully interpreted in the light of these variables. The total inability to respond to recall skin tests is called *anergy* and is generally an unfavorable prognostic sign.

Dinitrochlorobenzene (DNCB) Sensitization and Challenge

DNCB is a relatively simple chemical structure which is rarely encountered as a natural immunogen in the environment. Patients can be easily sensitized to the chemical by applying it dissolved in acetone to the skin. Patients may

95

then be challenged several weeks later with known quantities of the chemical and the dermal reaction can be quantitated. Usually patients are sensitized with 2000 μgm and rechallenged in 14 days with 25, 50, 100, and 150 μgm. Depending on the degree of reaction and at which dosage it occurs, the sensitization is quantitated and then followed serially during the course of the illness.

This type of immunologic evaluation tests both the afferent and efferent limbs of the immune response. That is, it tests the patient's ability to recognize a new antigen (afferent limb) and to mount an appropriate delayed hypersensitivity response when challenged (efferent limb). Thus, it adds an extra dimension to the information derived from the recall skin tests. In addition, the ability to quantitate precisely the challenge dose allows this parameter to be more easily interpreted. Data collected thus far has indicated a general correlation between the ability to become sensitized and a favorable prognosis. However, interpretations of this variable are also open to the criticisms that have been mentioned above under recall skin testing, i.e. the intensity of response may be modulated by a variety of circumstances not related to tumor immunity.

Evaluation of Peripheral Blood Mononuclear Cells

Another aspect of the immune system which has been extensively studied in cancer patients is peripheral mononuclear cells. Modern technology allows for the relatively easy separation of mononuclear cells from the remaining cellular constituents of the blood. B cells and T cells can be enumerated rather easily by the presence of surface immunoglobulin and the formation of sheep cell rosettes respectively. Monocytes can be distinguished morphologically and thus readily quantitated. The numbers of these various cells undergo major but inconsistent fluctuations in cancer patients possibly as a result of disease status and sometimes clearly as a result of chemotherapy or radiotherapy. As yet, no reproducible changes in the proportion or absolute numbers of these cells has been correlated with immunotherapy.

Perhaps more important than the numbers of the various cells, the functions of peripheral immunocompetent cells have also been extensively studied. An array of assays have been employed in hopes of obtaining a measure of the immune response which correlates most relevantly with the patient's clinical course. These efforts have been particularly intense in following patients receiving immunotherapy since there is the presumption that immunotherapeutic intervention ought to enhance a measurable parameter of immune function.

We as well as others have studied the proliferative response of peripheral blood mononuclear cells to a variety of stimuli. These stimuli have included polyclonal mitogens, such as phytohemagglutinin and conconavalin A, antigens, such as mumps antigen and PPD, and foreign cell surface antigens as in the mixed lymphocyte culture reaction (MLR). All of these functions measure the ability of peripheral blood mononuclear cells, lymphocytes for the most part, to divide in response to appropriate stimuli. With mitogens, for example, most normal lymphocytes will proliferate within a few days of exposure. With antigens, lymphocytes will proliferate only if the patient from whom they were isolated was previously sensitized to the antigen used. In the MLR, the patient's cells are exposed to foreign cells and proliferate in response to disparity within the major histocompatibility complex. All of these proliferative studies may be followed serially over a patient's course.

Finally, the cytotoxic potential of the peripheral blood mononuclear cells may be investigated serially. One assay, termed cell mediated lympholysis, is a two step assay. During the first step, the patient's cells are exposed and sensitized to allogeneic cells. During the second step, these now sensitized cells are exposed to identical ^{51}Cr labelled allogeneic cells. The sensitized cells may become "killer cells" and destroy the labelled allogeneic target cells. Cytotoxic *in vitro* assays would appear to be the most relevant to tumor immunology since they involve the recognition of cell surface antigens followed by the development of specific cytotoxicity against these antigens.

From a clinical point of view, these cellular assays, both proliferative and cytotoxic, have thus far been of limited usefulness. This is probably due to several factors. First, the innate fluctuations in the results of these assays make quantitation of responses and correlation with clinical status a frustrating task. Secondly, these assays do not measure tumor-specific immune responses. Reliable assays which measure tumor-specific immunity which have yet to be devised in humans hold the most promise of correlating immune status with clinical status. Finally, it is conceivable that no *in vitro* measurement of peripheral blood cells reflects the operation of the immune system *in vivo* and that other means of evaluating tumor immunity must be found.

6.0

Immunotherapy

Immunotherapeutic intervention in humans can be divided into four categories for purposes of discussion: local immunotherapy, non-specific immunopotentiation, specific active immunotherapy, and specific passive immunotherapy.

Local Immunotherapy

Local immunotherapy has been used for malignant melanoma and for other skin malignancies for about a decade. Unlike other forms of immunotherapy, local immunotherapy has a clear and reproducible effect in controlling malignant disease, albeit local control.

In general, patients with skin malignancies or soft tissue deposits of melanoma are made immune to BCG, DNCB, or PPD. The tumor mass is either painted, as in the case with

DNCB, or is inoculated with the appropriate antigen, e.g. Bacillus Calmette-Guérin (BCG). An immune response with its concomitant inflammatory component then occurs at the site of the inoculation and frequently the local tumor regresses. This tumor regression may reach the point of total local eradication of the lesion. Much less commonly, tumor masses in other areas may regress or completely disappear. This mode of therapy is locally effective in 55 to 90% of instances reported in the literature depending on the tumor type. The principal failing of this treatment is that it rarely, if ever, affects tumor deposits in the viscera.

A second form of local immunotherapy has been recently investigated in patients with Stage I carcinoma of the lung. Originally, it was observed that patients who had empyema complicating surgery for removal of lung cancer had a lesser chance of recurrence than patients who had uncomplicated postoperative courses. An investigation was then begun to determine whether the production of an empyema with intrapleural BCG would favorably affect the clinical outcome. Early reports from these investigations appear promising but more patient follow-up and confirmation of these findings are needed.

Non-specific Immunopotentiation

Non-specific systemic immunopotentiation is the presumed mode of action of a number of investigative agents in cancer therapy. These agents include Bacillus Calmette-Guérin (BCG), the methanol extracted residue (MER) of BCG, Corynebacterium parvum, and Levamisole. All of these agents are thought to enhance tumor-specific immunity as well. Their anti-tumor effect may be demonstrated under certain conditions in animals. These agents will now be discussed. Since the greatest experience has been accumulated with BCG, this agent will be discussed in greatest detail.

1. *BCG*

Bacillus Calmette-Guérin was originally developed as an attenuated Mycobacterium bovis for use in the vaccination against tuberculosis. Early animal studies suggested that there was a generalized immunostimulant effect from BCG and that under appropriate conditions an anti-tumor effect could be demonstrated. These observations have led to a large number of animal tumor and human tumor studies.

The most commonly practiced technique of systemic BCG application in patients is either by intradermal injection or by "scarification." Scarification (see Figure 6.1) is accomplished by scratching a "grid" on the skin with an 18 gauge needle approximately 4 cm by 4 cm on a previously cleaned area either on the upper arm or leg. Although this area may be preanesthetized with ethylene chloride, the discomfort is minimal even without anesthesia.

After capillary oozing occurs, the area is overlaid with a

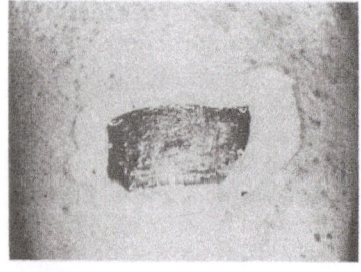

Figure 6.1
The application of BCG by scarification necessitates interruption of skin integrity and the development of a well of vaseline within which the BCG material is placed.

culture of BCG organisms (usually $3-8 \times 10^8$ organisms)
and the organisms are allowed to absorb locally. The organisms usually form a local lesion similar to but much
less destructive than the classical "cold abscess" of tuberculosis. The organisms survive locally for many months
and frequently spread, usually via the lymphatics, in a
contiguous fashion. In most human studies patients are
scarified either weekly or semi-weekly until sensitization
occurs as indicated by a PPD skin test response. This sensitization can then be maintained by less frequent scari-
fications. Overall, the treatment is well tolerated and only
rarely do patients refuse to continue therapy.

The toxicity of BCG therapy was more substantial in the
early human studies. This result is probably due to the
fact that many patients received BCG during the terminal
phases of their illnesses. Terminally ill patients frequently
have markedly diminished host defenses and BCG becomes more virulent under these conditions. Thus, gran-
ulomatous hepatitis, BCG meningitis, and an illness sim-
ulating disseminated tuberculosis were observed. This
group of infections responded to anti-tuberculous therapy
but was alarming to some physicians.

In more recent studies, BCG has been used to treat patients who are basically well with either minimal residual
disease or no evident disease but with a high likelihood of
recurrence. In this group of patients the principal side effects of BCG are relatively minor and include fever, a flu-
like illness, and lymph node enlargement. As with all
agents, and especially immunotherapeutic agents, severe
allergic reactions with anaphylaxis can occur but are rare.
We have not observed severe complications in relatively
well patients using the scarification technique.

The results of systemic BCG treatment in patients are still
not entirely clear. Many of the studies in melanoma and
leukemia, for example, which show a beneficial effect of
BCG therapy, suffer from serious flaws of experimental
design. However, from animal studies a number of important guidelines to direct future efforts have been determined:

    a) there must be a relatively small tumor burden or the
       immune system is overwhelmed;

    b) the contact between BCG and the tumor cells should
       be as direct as possible;

    c) the host must have the ability to respond to the my-
       cobacterial antigens presented if any anti-tumor re-
       sponse is expected.

The inclusion of these guidelines in future studies should
clarify the usefulness of BCG as a clinical immunopotentiator.

2. *MER*

MER is the methanol extracted residue of a killed culture
of BCG organisms. It was developed from BCG because of
the desire for a nonviable anti-tuberculous vaccine. MER

has been considered to be the "active principle" of BCG and has been used as an immunopotentiator in cancer patients.

MER may be quantified and is usually administered as multiple intradermal injections (Figure 6.2). The injection sites frequently become inflamed, especially as sensitization proceeds. If the injections are too deep or the dose too high, sterile abscesses result which may drain purulent material for a period of months. The toxicity of MER is primarily local from the discomfort in the area of the injection sites. Patients generally need support to complete a course of therapy. However, the systemic toxicities of MER are minimal with fever and lymph node enlargement being the only frequently encountered complications. Since MER is nonviable, none of the infectious problems encountered with BCG are seen.

MER is presently being studied intensively as a non-specific immunopotentiator in a variety of clinical settings. There is some enthusiasm concerning its possible efficacy in the treatment of acute myelogenous leukemia and colon cancer but the studies are far too preliminary to assess its role.

3. *Corynebacterium Parvum*

C. parvum is a gram positive anaerobic bacterium which was found to enhance host defenses in animals when injected systemically and has been utilized in a variety of clinical cancer studies. The organism is usually administered in a killed vaccine and has thus far been used principally as an adjunct to chemotherapy to boost immune responses. Critical reports are sparse but meaningful studies are in progress.

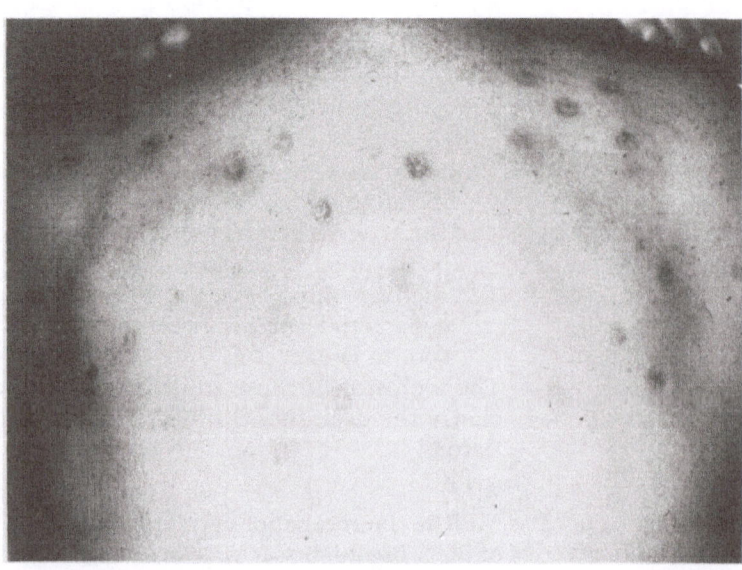

Figure 6.2
Intradermal application of MER-BCG is an alternative to scarification and may result in dermal ulceration following the local inflammatory response.

4. Levamisole

Levamisole is an antihelminth drug which was found ser-
endipitously to enhance delayed hypersensitivity skin
tests in patients taking the drug for parasitic infections.
The mechanism of this action is not clear and whether or
not it represents a true immunologic effect is unclear. Its
principal attribute is that it may be taken orally. This drug
is being studied extensively in trials with cancer patients.

Specific Active Immunotherapy

Specific active immunotherapy refers to the immunization
of patients with tumor antigens in attempts to either protect
them from developing these tumors or in hopes of boosting
the endogenous response to a tumor already present. This
therapy may be approached from three aspects.

1. Immunization with Tumor-Virus Vaccines

Vaccination with viruses as a means of preventing viral
infectious diseases such as smallpox and polio is clearly
an effective maneuver if the appropriate procedures can be
developed. The prevention of cancer in humans via this
technique is a theoretical possibility if vaccination against
human oncogenic viruses can be achieved. Vaccination
has been accomplished in at least one animal malignancy,
i.e. Marek's lymphomatosis in chickens. The full development of this area must await advances in the area of
viral oncology.

2. Immunization with Tumor Cells

TSTA presumably exist on the surface of tumor cells and
therefore it should be possible to utilize killed tumor cells
to sensitize patients against tumor antigens. This strategy
has been employed in both animal and human studies
with variable success. In humans, patients have been inoculated with autologous tumor cells or allogeneic tumor
cells. Additional studies have attempted to increase the
immunogenicity of the tumor cells either by mixing them
with non-specific immunopotentiators such as BCG or by
treating the cell surface with neuraminidase. The human
studies are still in progress.

3. Immunization with Tumor Specific Transplantation An-
tigens

Exploration of this area will require isolation of and probably purification of TSTA in humans. This requirement
has not yet been accomplished. Thus, this strategy must
await further technological advances.

Specific Passive Immunotherapy

Passive immunotherapy refers to the use of the end products
of the immune response as, for example, antibodies or sensitized cells, passively transferred to patients in an attempt
to perturb the tumor. For instance, patients who are cured
of locally extensive tumors such as breast cancer are as-

sumed to have particularly effective host defenses against that tumor. Part of this assumed host defense may be the presence of anti-tumor antibodies in the serum or cells sensitized against tumor antigens. On these assumptions clinicians have infused serum from one patient to another, cells from one patient to another, and a factor called transfer factor from one patient to another with therapeutic intent.

1. Passive Infusion of Anti-Tumor Serum

Reports of the passive infusion of anti-tumor antibodies in the form of serum from patients who have had favorable responses do not stand up to critical analysis. Most importantly there is no available system by which one can detect anti-tumor antibodies and, thus, it is impossible to interpret the results. Since it is never clear whether any anti-tumor antibody has been administered, it is not apparent whether one is viewing a response or a random fluctuation in the natural history of an illness. When reliable systems become available to detect and quantitate anti-tumor antibodies, this modality of treatment should be evaluated.

2. Passive Infusion of Transfer Factor

Transfer factor is a low molecular weight nucleoprotein which can transfer "cellular immune reactivity" from one individual to another. The mechanism of action is entirely unclear. However, it is easily prepared and has been used to advantage in certain immunodeficient patients. Transfer factor has been administered to cancer patients with anecdotal success. However, inability to determine a "sensitized" transfer factor donor because of the technical problems with detection of human TSTA have hindered research in this area.

3. Passive Infusion of Sensitized Cells

Since sensitized cells are believed to be the most important factor in tumor immunity, it seems reasonable to explore the therapeutic potential of sensitized lymphocytes. Studies in this area have once again been hindered by difficulties in defining tumor-specific immunity. However, the exquisite specificity and efficient cytotoxicity of "killer" cells *in vitro* suggests that this area of research may be quite fruitful and perhaps ultimately yield the "magic bullet" which can distinguish and selectively destroy tumors.

7.0   **Future Considerations**

Despite a somewhat tentative beginning, the relationship between immunology and cancer on the clinical level should become more productive. A number of avenues of research have now been established at research institutes. Investigators are exploring multimodality approaches in which non-specific immunotherapy is combined with either chemotherapy or radiotherapy to alleviate the inherent immunosuppressive effects of chemotherapy and radiotherapy. Other investigators are exploring combinations of immunotherapy such as the use of autologous blast cells and MER in leukemic patients. Several other groups are trying to define the precise mechanism of action of immunotherapeutic agents so that the immune system may be manipulated from a standpoint of understanding rather than of empiricism.

However, the most important future contributions in immunotherapy will almost certainly be the result of advances in laboratory research. The immune system is an extraordinarily complex system which is by no means fully understood. Further insights into the science of this fascinating system may allow us to harness its enormous specificity for clinical diagnostic and therapeutic advantage.

References

Oettgen, H. F., Immunotherapy of Cancer. *New Engl. J. Med.* **297**:484–492, 1977.

Burnet, F. M., The Concept of Immunological Surveillance. *Prog. Exp. Tumor. Res.,* **13**:1–27, 1970.

Bast, R. C., Jr., Zbar, B., Borsos, T., and Rapp, H. J., BCG and Cancer. *New Engl. J. Med.* **290**:1413–1420, 1458–1469, 1974.

Nathanson, L., Use of BCG in the Treatment of Human Neoplasms: A Review. *Sem. in Oncol.* **1**:337–350, 1974.

Powles, R. L., Immunotherapy in the Management of Acute Leukemia. *Brit. J. Haem.* **32**:145–149, 1976.

Section III

Management of Cancer Signs and Symptoms

7

Clinical Management of Marrow Suppression and Immune Suppression

1.0

Introduction

Marrow suppression and immune suppression commonly accompany cancer. They are most frequently observed in the leukemias and other diseases associated with replacement of the marrow and in primary malignant diseases of the immune system. In addition "solid" tumors or non-hematologic malignancy may metastasize to the bone marrow replacing normal elements. Finally, treatment directed at the tumor may incidentally affect production of cells. Bone marrow suppression is associated with the secondary development of anemia, leukopenia, or thrombocytopenia. Peripheral and central depletion of the "formed elements" of the blood—that is, the red blood cells, white blood cells, and platelets—leads to anemia, and can bring about infection and/or bleeding. Immune suppression may be distinguished from marrow suppression in that the cells involved in the immune system are lymphocytes and plasma cells. Furthermore, the depletion of these elements may only be evaluated by specific tests, such as immunoglobulin analysis and the lymphocyte function tests. The management of marrow suppression and immune suppression is one of the most delicate clinical issues because the common causes of mortality in cancer result from secondary sepsis or hemorrhage. This chapter will focus on the management of anemia, leukopenia, and thrombocytopenia in the patient with neoplastic disease and also on the management of chemotherapy in patients with secondary marrow and immune suppression.

2.0

Differential Diagnosis of Pancytopenia

Pancytopenia or combined leukopenia, anemia, and thrombocytopenia in patients with underlying neoplastic disease may develop as a consequence of many factors. The two primary causative factors are drug induced marrow suppression and marrow suppression secondary to replacement by tumor. A number of solid tumors are associated with marrow metas-

107

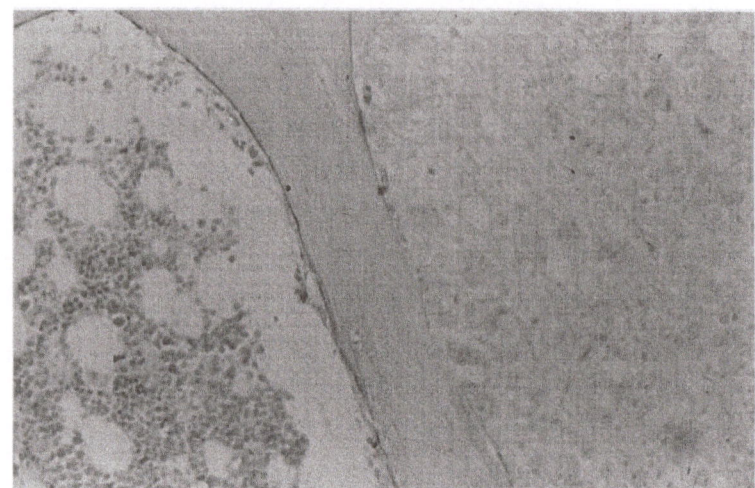

Figure 7.1
Bone marrow biopsy demonstrating normal distribution of marrow elements on the *left*. A spicule of bone separates the marrow compartment of the *right* which is invaded with small islands of tumor cells and replaced by a secondary scirrhous fibrosis reaction.

tases, particularly lung, breast, and prostate cancer. In addition, the leukemias, multiple myelomas, and non-Hodgkin's lymphoma are associated with primary marrow replacement. In those patients with marrow invasion secondary to solid tumor metastases, two clues to the presence of such a tumor in the marrow are: (1) radiographically demonstrable bone lesion by bone scan or metastatic series and (2) the presence in the peripheral blood of abnormally shaped (tear shaped) red blood cells as well as immature white blood cells and red blood cells known as normoblasts. This so-called leukoerythroblastic blood picture is typical and pathognomonic for an irritated marrow. Bone marrow biopsy will generally demonstrate tumor cells with or without normal marrow elements (Figure 7.1). Another manifestation of bone marrow invasion within the tumor is the development of secondary fibrosis (Figure 7.2). The microscopic appearance of the tumor cells is a syncitial coalescence of cells often appearing necrotic (Figure 7.3). De-

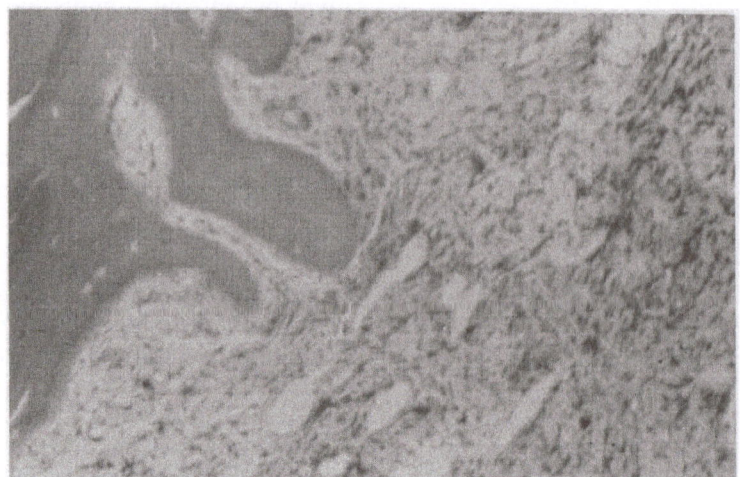

Figure 7.2
This bone marrow spicule is completely effaced by a fibrotic reaction which occupies the entire cavity; no tumor cells were identified. This is an example of primary myelofibrosis.

108

Figure 7.3
The marrow cavity pictured here is basically empty of normal marrow elements. The cellular debris observed is primarily stromal supporting structure for the marrow. This picture may be secondary to radiotherapy or to chemotherapy.

pletion of bone marrow elements by drug therapy or radiation results in the replacement of the marrow space with fat (Figure 7.4).

Another cause of pancytopenia or a decrease in each of the formed elements in the peripheral blood is hypersplenism. In patients with enlarged spleens, secondary to neoplastic disease, and generally as a consequence of liver involvement with increased portal pressure and secondary splenomegaly, the sequestration of cells within the spleen and the removal from the circulation of senescent cells results in a decrease in any or all of the formed elements. Isolated or solitary leukopenia, anemia, or thrombocytopenia may occur with hypersplenism although more commonly all three elements are modestly decreased. In hypersplenic patients bone marrow biopsy generally reveals hyperplasia of the normal marrow constituents in distinct contrast to the pancytopenia secondary to drug or radiation therapy or tumor marrow replacement.

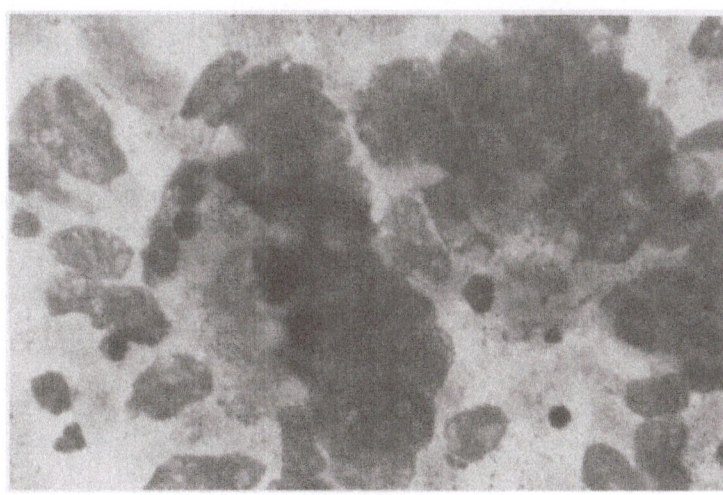

Figure 7.4
Tumor cells in the marrow appear under high power as several clumps in a syncitial arrangement. The cells may be necrotic, blurring the cell wall details.

A discussion of the differential diagnosis of the isolated or solitary anemia, leukopenia, or thrombocytopenia is detailed within the respective sections.

3.0   **Chemotherapy-Related Marrow Suppression**
The use of cytotoxic drugs in the treatment of cancer has become an important part of the total care concept for the cancer patient and has added substantial impetus to the prospects for cure in a number of hematologic as well as solid tumors. A characteristic of the majority of cancer chemotherapy agents is the lack of tumor specificity. Consequently, the drug effects on normal host tissues and particularly upon the rapidly dividing cells within the bone marrow are expected and occur regularly. The marrow suppression secondary to cytotoxic drugs represents the basic dose limiting factor for the majority of the active tumor agents. Furthermore, an essential principle of drug therapy of cancer is that a dose response relationship exists with regard to the anti-tumor response observed. Marrow suppression, therefore, is not only a common accompaniment of therapy but is important as a monitor or signal to ensure that an effective drug level is achieved. The dose response issue can be described in this way: when the dose is increased, a greater number of tumor cells are killed and there is a greater likelihood of an anti-tumor effect. However, a maximum point exists beyond which the adverse effects of the drugs on normal host cells limit the dose. Exceeding the acute dose level may result in adverse host effects and possibly drug induced mortality.

The clinical management of drug induced secondary anemia, leukopenia, and thrombocytopenia is an essential part of the cytotoxic chemotherapy of neoplastic disease and involves careful patient monitoring, supportive care and the prudent use of blood product transfusion. In addition to diagnosis or detection of marrow suppression, such effects can be anticipated by identifying those risk factors which may predict the secondary development of pancytopenia. For almost all chemotherapeutic agents the development of marrow suppression is related primarily to the dose of the drug administered. Host factors may also alter the metabolism, distribution, and pharmacology of a drug and may result in additive adverse effects on the bone marrow. The major factors to be considered in predicting potential adverse drug pharmacology and secondary "cytopenias" are listed in Table 7.1. Performance status defines the patient's functional status in relationship to the patient's ability to ambulate and is a gross estimate of overall health. A correlation of the tumor response to chemotherapy with pretreatment performance status is well established. Certainly patients who are cachectic and bedridden with a performance status of 4 have less tolerance to drug therapy. The drugs are therefore less likely to have an anti-tumor effect and to prolong survival.

In addition to performance status, prior exposure to che-

Table 7.1
Factors Determining Development of Marrow Suppression with Cytotoxic Drugs

Drug Factors

1. Class (alkylating agent, antimetabolite, etc.)

2. Dosage and schedule

Host Factors

1. Prior systemic (chemo) therapy or radiotherapy

2. Performance status

3. Bone or bone marrow disease

4. Age

5. Liver function

6. Renal function

motherapy or radiotherapy may be considered the major determinant of the adverse effects of drugs on the marrow. Previous exposure to chemotherapy or radiotherapy compromises the stem cell elements and the "bone marrow reserves" resulting in limited ability to deliver the usual doses of drugs. Bone metastases or marrow invasion are additional major factors in predicting marrow suppression. For patients with marrow tumor replacement of stem cells and normal elements, the dose of drug therapy may be limited because of the decreased marrow reserve. The therapeutic intention, however, is to induce sufficient tumor regression to "unpack" the marrow and to allow normal marrow elements to grow. In this circumstance, in patients with an exquisitely sensitive tumor, one may initiate therapy with low doses of drugs and increase the dose as the marrow is reconstituted with normal cells. Alternatively, patients with less sensitive tumors may require higher doses of drugs to clear the marrow of tumor. This may deplete the normal stem cells and preclude prompt reconstitution of peripheral blood counts. Thus, the paradox of employing a high dose of drugs in patients with limited marrow reserve is a unique but rational therapeutic approach.

A disturbance of chemotherapy drug pharmacology may result from hepatic or renal function abnormalities secondary to the tumor or from associated diseases, such as nephritis or cirrhosis, which in turn results in prolonged circulating blood levels of drugs. The longer duration of marrow exposure results in a high-dose effect on the marrow. Some chemotherapy drugs may require activation in the liver. The inability to convert the drug to the active form will result in a lesser drug effect for patients with compromised liver function. Thus, for cyclo-phosphamide which requires metabolic activation by the liver, an abnormality of liver function may result in decreased drug effect. Another example of the importance of hepatic function can be observed in patients treated with adriamycin, a drug which is secreted via the biliary tree into the bowel. In patients with biliary obstruction, prolonged blood levels of adriamycin may be observed resulting in excessive toxicity.

There are also drugs which are excreted primarily by the kidneys. In patients with abnormal renal function related to obstruction or chronic intrinsic renal disease, the blood levels may be prolonged. The result of the prolonged levels is the development of major adverse effects on the host, unless the drug is metabolized and excreted. In the case of methotrexate, which is excreted within 24 hours of drug delivery but is not metabolized, the presence of abnormal renal function results in major toxicity.

There are five classes of chemotherapeutic agents outlined in Chapter 5. All of these agents with a few exceptions cause bone marrow suppression which develops in a standard chron-ologic pattern. Five individual agents do not generally affect the marrow when the drugs are employed in optimal thera-peutic dosage. In patients with a limited marrow reserve, how-ever, even these drugs may be associated with pancytopenia. Depending on the class of drugs as well as the host factors mentioned previously, peripheral blood counts begin to fall from 4 to 14 days after treatment with the nadir or low point reached on day 8 to 16. The drugs which are notable exceptions to this pattern of marrow suppression are the nitrosourea com-pounds which characteristically lead to delayed marrow suppression with the nadir developing on day 28 in a range from 14 to 40 days. A summary of chronologic patterns of leukopenia and thrombopenia is detailed in Table 7.2.

Recovery of marrow function may be delayed according to a number of factors including marrow reserve but generally recovery is apparent in 5 to 7 days. Cumulative marrow suppression—that is, increased marrow effects with repeat drug exposure—is characteristic of the nitrosourea compounds and possibly for the other alkylating drugs; but the majority of chemotherapy drugs cause transient and completely irre-versible marrow suppression.

An interesting phenomenon is the selectivity of some drugs in affecting one of the formed elements (leukocytes, erythro-cytes, and platelets) rather than all three. Cyclophosphamide primarily affects the white blood cells and is platelet sparing. Alternatively, the nitrosoureas cause thrombocytopenia while leukopenia is quantitatively of lesser significance. The general sequence of events in the development of pancytopenia is to observe leukopenia followed in days by thrombocytopenia and generally weeks later by anemia. The recovery of blood counts proceeds in the same sequence and is often accompanied by a rebound leukocytosis or thrombocytosis. This phenomenon of rebound or overshoot is most commonly observed in patients

Table 7.2
Schematic Approach to the Patient with Marrow Suppression

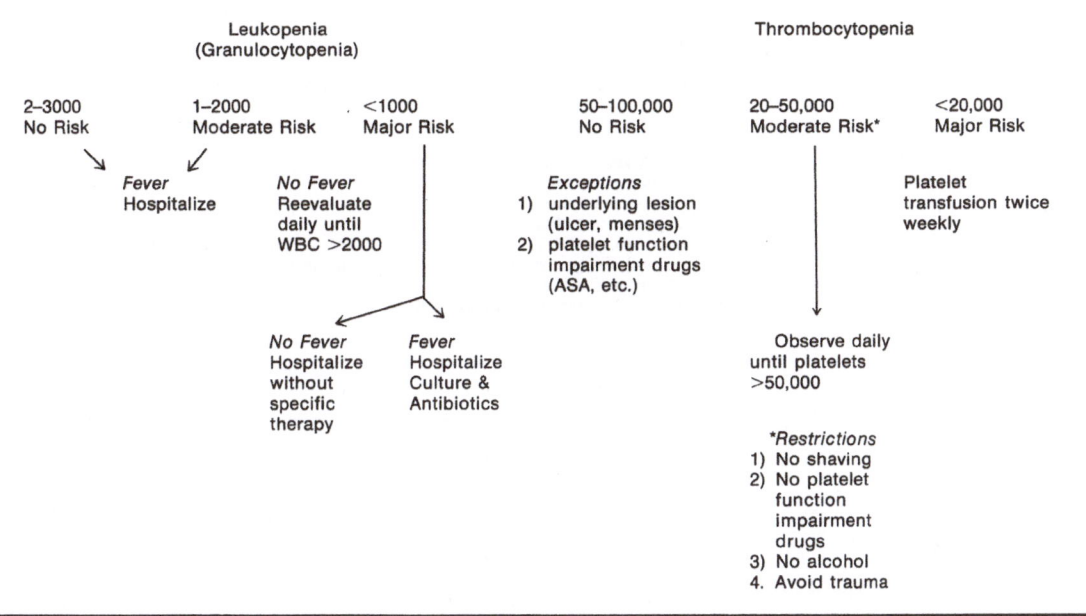

treated with anti-metabolites but may be seen in other treatment programs as well.

The question of specific management of cytotoxic drug therapy for patients developing or potentially developing marrow suppression may be approached in a variety of ways. Some of the general principles involved in initiating therapy and adjusting the drug dosage are outlined in Table 7.3. For most drug treatment programs involving intermittent therapy, drug doses may be adjusted on the basis of the nadir blood counts. Therapy may also be interrupted for patients whose blood counts are compromised at the time of the next course: the schedule as opposed to the dose is altered. For patients receiving continuous or daily drug therapy, the drug is interrupted until recovery is evident. See Table 7.4 for guidance on how to monitor blood counts.

4.0

Radiation Induced Marrow Suppression
Unlike chemotherapy which is systemic in distribution and therefore affects all functioning marrow, radiation therapy is administered as local therapy and affects only that marrow which is exposed to the treatment beam. As with drug therapy, the effect on the marrow is dose related but in radiation therapy, volume of treatment is important and determines the potential effects of radiation on the marrow. In normal patients the pelvis accounts for approximately 35% of the total functioning marrow and the vertebral column an additional 35%. The final 30% is accounted for in the proximal ends of the long bones and in the ribs. With increasing age the marrow

Table 7.3
General Principles of Cytotoxic Drug Administration

1. Initiate therapy with 50% of optimal dose or protocol doses if patient:

a) is >65 years of age
b) has had prior radiotherapy
c) has >2 defined bone lesions
d) has had prior chemotherapy
e) has major systemic symptoms (weight loss >10% or temp >101°)
f) has major organ dysfunction which would affect drug pharmacology—bilirubin >3; creatinine >1.5
g) has WBC <3000 or platelets <100,000

2. Adjustment of drug dosage

a) escalate or de-escalate *only* the myelosuppressive drugs
b) adjustment of dose based on IMMEDIATE pretreatment WBC and/or platelets:

if WBC <3000 or platelets <100,000: no therapy for 1 week

if WBC 3000–4500 or platelets 100–125,000: administer 50% doses

c) adjustment of dose based on NADIR level with normal WBC (>4500) and platelets (>150,000) at time of treatment:

WBC or	Platelet	Drug Dose
<1000	<50,000	50%*
1000–2000	50,000–100,000	100%*
2000–3000	100,000–150,000	125%*

d) adjust dosage of *all* myelosuppressive drugs in a combination program

*% refers to decrease or increase of drug dosage administered in prior course of treatment

distribution changes in two ways: the marrow may be replaced in part by fatty tissue and the functioning marrow shifts to the long bones. For patients receiving radiation therapy to doses beyond 3500 rad, the marrow effect may be irreversible as demonstrated by bone biopsies done as much as 2 years following radiation.

An interesting phenomenon is that leukopenia or thrombocytopenia may develop after radiation of small volumes. This development occurs particularly after radiation of the thymus or the spleen. The radiation most likely causes the release of some humoral factor which secondarily affects the bone marrow. This so-called abscopal effect has been observed most commonly in chronic myelogenous leukemia therapy.

5.0

Anemia
There are often many causes of anemia in patients treated with cancer chemotherapeutic agents. For this reason, a search for

Table 7.4
Guide to Monitoring Blood Counts

1. Adjust initial dose based on host and tumor factors

(see Table 1).

2. Monitor WBC, HCT, and platelet count weekly.

3. Nadir generally observed

d 7 Antimetabolites and alkylating agents

d 14 Antibiotics (specifically anthracyclines)

d 21 (Nitrosourea or Mitomycin C only)

4. Nadir level: general approach (see Schematic)

a) For early (d 7) leukopenia or thrombocytopenia (<4000 or <150,000) monitor more frequently (every 2 days).

b) For nadir levels consistent with no significant risk patient can be seen at weekly intervals.

c) For nadir levels associated with moderate risk (WBC 1–2000 or platelets 20–50,000) patient should be seen daily.

d) For nadir day levels associated with major risk, patients should be hospitalized *or* observed closely for symptoms or signs of secondary effects.

5. Consecutive day levels (CDL) interpretation

a) WBC and platelet levels obtained on consecutive days are useful in that the rate of fall may help predict the nadir level as well as the nadir time.

b) CDL which are the same *or* within a single standard error generally imply that the nadir has been reached and no further fall is anticipated.

c) CDL Platelets: the daily incremental decrease in platelet count in progressive thrombocytopenia is 20%.

Example: Patient received chemotherapy on day 1. Platelet count on day 10 is 80,000 and on day 11 is 60,000. Prediction for major risk thrombocytopenia is on day 13 or 14.

d) CDL White Blood Cells: the daily incremental decrease in WBC count in progressive leukopenia is not as predictable. The increment change may be sudden and severe, e.g. 2500 to 500 over a single day.

e) Duration of myelosuppression (leukopenia or thrombocytopenia) is generally 5 to 9 days with overshoot or rebound leukocytosis or thrombocytosis commonly observed.

non-drug related causes should be pursued, particularly when anemia is unassociated with leukopenia or thrombocytopenia. Gastrointestinal bleeding, occult or with melena, is the most common cause of anemia in the cancer patient. The nonspecific anemia of chronic disease is secondary to ineffective iron utilization. This anemia is a common condition and may be distinguished from blood loss by evaluation of marrow iron

115

stores. Hemolysis may be specifically associated with some malignant disease such as chronic lymphatic leukemia, but is relatively infrequent as a cause of anemia in malignancy. The anemia induced by chemotherapy is secondary to decreased red blood cell production and generally develops as a cumulative effect of multiple courses of treatment and is almost always reversible with drug withdrawal.

Transfusion of red blood cells is often salutory in patients with anemia but should be reserved only for patients with significant or major symptomatology. The symptomatic indications for transfusion therapy of drug-induced anemia are variable, but the most common indication for transfusion is the constitutional syndrome of weakness and lassitude. This nonspecific symptom complex may be related to other causes in the cancer patient and unless the hematocrit is less than 25%, alternative causes of such symptoms should be sought, such as hypokalemia or hypercalcemia. Specific symptoms such as dyspnea or angina are only rarely related directly to anemia; however, anemia with hematocrit less than 20% in association with underlying pulmonary disease or coronary artery disease may result in major symptomatology. In such circumstances transfusion therapy is not only indicated but is essential.

The potential adverse effects of transfusion are hepatitis, hypersensitivity reaction or sensitization to the formed elements, and volume overload. One must balance such effects against the fact that drug induced anemia is generally transient and spontaneously reversible with drug withdrawal. It is clear that RBC transfusion should be employed cautiously, prudently, and probably infrequently. When used, packed red blood cells (PC) should always be employed in preference to whole blood to minimize the volume infused and the potential for transmission of hepatitis. Washed or fresh frozen red blood cells should be used in patients sensitized to multiple HLA or white blood cell antigens which are present in standard transfusion packs, or in patients (transplantation candidates) in whom sensitization should be avoided.

Supplemental vitamins such as folic acid or vitamin B_{12} may be useful especially for patients with prior gastrectomy and are not contraindicated by the anti-tumor chemotherapy. Iron therapy in patients with blood loss episodes should be employed, particularly when the blood loss is not completely corrected with transfusion. Infrequently recognized sources of blood loss are menstrual loss in females and depletion secondary to routine daily blood letting for biochemical assays.

6.0
Leukopenia

Leukopenia is the most common drug-induced effect of marrow suppression. The leukocyte population is composed of two components: lymphocytes and granulocytes. Lymphocytopenia is the earliest effect of anti-tumor drugs and is most commonly observed with alkylating agents and corticosteroids. The peripheral blood lympholysis leads to immune suppression and secondary predisposition to opportunistic infections. Granulocytopenia or neutropenia develops in se-

116

quence following lymphopenia and is the major risk factor for both opportunistic and common bacterial pathogens. Reconstitution of circulating white blood cells develops in 3 to 7 days following the nadir and is heralded by a monocytosis, followed by an increasing percentage of granulocytes daily in the peripheral blood.

The general strategy of monitoring patients during the myelosuppressive period is to anticipate the nadir of granulocytopenia. In general, leukopenia is reversible but with blood counts (WBC) <1000 or granulocytes <200 cells/mm³, the risk of sepsis is major. Hospitalization and sterile precautions, such as isolation of the patient, are mandatory particularly if the duration of neutropenia is to exceed 3 days. The reason for hospitalization is that infection may develop suddenly with fever, sepsis, and shock arising in the space of hours and necessitating prompt therapeutic intervention. In patients with nadir granulocyte counts above 200 cells/mm³, the risk of infection is increased but no specific therapy is required and hospitalization is not necessary.

Antibiotics generally should not be administered prophylactically but with the development of fever, all orifices or potential sites of infection should be cultured and broad spectrum antibiotic coverage should be initiated. The specific choice of antibiotics is controversial. Gentamycin is almost always employed because of its gram negative spectrum; carbenicillin is added because of the risk of pseudomonas infection; a cephalosporin is included for the gram positive organisms; and finally, clindamycin may be added if anaerobic organisms are suspected. The specific agent or combination will always depend upon an intensive search for the offending organisms. The most common source for such infections is the gastrointestinal tract because of the common development of mucosal ulcerations induced by the drugs.

Granulocyte transfusions have become a practical reality as a consequence of developments in recent technology. Granulocytes from patients with chronic myelogenous leukemia or from normal donors are generally available only at major research centers. Harvesting of WBC is achieved by two methods: centrifugation and filtration which are generally equal with regard to the yield of the numbers of cells and the viability and functionality of the cells. Two studies have demonstrated the effectiveness of granulocyte transfusions in reducing fever and prolonging survival in patients with marrow suppression, but the general application of transfusion remains controversial. The specific indications for white blood cell transfusion are as follows: (1) severe granulocytopenia (<200 cells/mm³); (2) clinical sepsis as determined by fever and a positive culture source; and (3) recovery from marrow suppression not anticipated for more than three days. In general, only patients with all three indications should be considered for white blood cell transfusion. The restricted availability of white blood cells for transfusion and the inconclusive data on effectiveness makes such transfusion experimental.

Isolation and the use of protected environment, either prophylactically or therapeutically, may be effective in reducing the frequency of infection in leukopenic patients. This concept, however, is still controversial. Although neutropenic patients without fever should not be exposed to other patients with possible infections, they need not be isolated from contact with the health care team and their family. The patient with fever and infection need not be isolated since the presence of the infection is established and not necessarily transmissible to staff. "Precautions" protect the staff or visitors but do not protect the patient and the effect of isolation is detrimental to the necessary standards of patient care and to the patient's psychological stability.

Prophylaxis in the prevention of infection in patients receiving chemotherapy should involve a bowel sterilization with non-absorbable antibiotics including anti fungal and bacterial drugs and antiseptic care of mucosal saprophytes such as the oral, vaginal, and peri-anal areas. Procedures which lead to transient bacteria, such as teeth brushing and barium enema, should be carefully avoided. Interruption of cutaneous integrity, particularly by venapuncture, is also to be minimized.

7.0 Thrombocytopenia

Thrombocytopenia almost never occurs as an isolated clinical observation, but is usually seen in association with moderate to severe leukopenia. Spurious thrombocytopenia and idiopathic thrombocytopenia secondary to immunologic mechanisms should be investigated as alternative causes of low platelet counts in patients on chemotherapy. Another rare but important cause of a decreased platelet count is disseminated intravascular coagulation (DIC) which may develop in association with infection and commonly as a part of the general process of neoplasia.

Thrombocytopenia secondary to cytotoxic drug effects is almost always reversible but hemorrhage is a major risk in such patients and the second most common cause of death in leukemic patients is thrombocytopenic induced hemorrhage. Spontaneous bleeding never occurs at platelet levels above 50,000 cells/mm^3 and hemorrhage is only a significant risk when the platelet count is less than 20,000 cells/mm^3. Complicating factors such as trauma, severe sunburn, second degree burn or the use of medications known to interfere with platelet function, such as alcohol, aspirin, and anticoagulants, may result in major hemorrhage at platelet levels over 20,000 cells/mm^3 because of the additive insult to the hemostatic mechanism. When women menstruate, suppressive estrogen therapy is necessary to avoid profound uterine hemorrhage secondary to thrombocytopenia at the time of normal menses. Thus, a standard procedure in female patients being treated for leukemia is to place them on oral contraceptives to suppress menstrual flow and to avoid thrombocytopenia-induced secondary hemorrhage. In addition to instructing patients with regard to platelet interfering drugs, it is important to emphasize the

118

avoidance of trauma. An often unrecognized self-induced trauma is the Valsalva maneuver. This maneuver is commonly a part of daily activities such as stool straining or picking up heavy objects. In both instances, the result is increased intra-abdominal pressure and intra-cerebral pressure. Secondary hemorrhage, particularly intra-cerebral, is a major risk in this instance.

Chronic thrombocytopenia secondary to drug therapy is uncommon but patients may tolerate prolonged periods of thrombocytopenia even to levels of 10–15,000 cells/mm³ without bleeding. Such patients should be monitored for the development of clinical occult bleeding. More importantly, the additional insult of infection may potentiate the thrombocytopenia and result in secondary hemorrhage.

The major concern for patients with thrombocytopenia is the development of critical site hemorrhage, particularly intracerebral or gastrointestinal. In almost all instances, hemorrhage from mucosal or visceral sites, such as the gastrointestinal tract, kidney, and brain, is preceded by the characteristic "rash" or petechial hemorrhage into the skin. Platelet transfusion replacement therapy may be reserved for symptomatic or recognized petechial bleeding. However, acute anemia, sepsis, shock, or unusual central nervous system symptoms in thrombocytopenic patients may be clinical clues to major hemorrhage even in the absence of cutaneous bleeding. Platelet transfusions should be employed in these instances.

The transfusion of platelet-rich concentrates is now standard practice and is available through the services of most blood banks. In general, platelet transfusion should be reserved for patients with active skin or mucosal bleeding who have a platelet count less than 50,000 cells/mm³. The strategy in monitoring thrombocytopenia secondary to drug therapy is similar to monitoring for leukopenia: one must anticipate the platelet nadir and institute specific therapy, platelet transfusion, when complications develop. Prophylactic platelet transfusion has been advocated for some patients with platelet counts less than 20,000 and if a prolonged period of thrombocytopenia is anticipated.

Specific guidelines for the administration of platelets depend upon the response to transfusion—that is, the platelet increment—and the clinical control of bleeding. HLA type specific matched platelets are more effective than random donor platelets in terms of incremental increase in the platelet count and the platelet life span in the circulation. In general, 4 to 8 units of random donor platelets should be administered or 2 to 4 units of HLA type specific until hemorrhage is controlled. HLA type specific platelets are only used in patients refractory to random donor platelets and are only rarely necessary, except for patients receiving frequent transfusion leading to sensitization. In patients with disruption of mucosal integrity, such as a duodenal ulcer or a bleeding lesion secondary to a vascular break, a surgical control is paramount in conjunction with simultaneous platelet transfusion. Finally, platelet transfusion is absolutely indicated for patients with sepsis in whom platelets may function to augment the leukocyte phagocytic system and other components of the host's defense mechanisms.

Immune Function Abnormalities: Detection and Management

Immune function has been reviewed in Chapter 6 in relation to the various immune therapies applied to the cancer patient. This function is still an area of major clinical investigation and experimentation. It is well known that the host's immune function is adversely affected by neoplastic disease as well as by the therapy directed at the disease. For example, radiation therapy directed at the chest and thymus area results in decreased T cell function, but the clinical effect on the disease or on the host is unclear.

Immunologic reconstitution by transfusion, transplantation, or antibody restitution is experimental at the present time. As yet, no form of immune therapy, either stimulation, potentiation or replacement, has been clinically useful. Clinical trials in malignant melanoma, breast cancer, acute leukemia, and many other tumors are ongoing. A major goal of such studies is to develop ways of monitoring the effect of immunologic treatment on the tumor as well as on the immune system because, at least in theory, immune therapy could result in enhancement of tumor growth. The development of opportunistic infection, by which is meant unusual infections secondary to fungi, viruses, or even protozoa, is presumably related to a compromise in the immune system.

The awareness that anti-tumor therapy causes secondary effects on the immune system has resulted in different methods of treatment delivery. For example, intermittent drug treatment schedules, permitting immunologic reconstitution between courses of therapy, are commonly employed.

Table 7.5
Principles of Replacement Therapy for Blood Products in Depletion Secondary to Drug Therapy with Underproduction

Blood Product	Indication for Use	Quantity and Frequency	Comment or Ancillary Procedures
Packed Red Blood Cells	Hct 25% with symptoms	2 units	Folic acid, B12, and iron if indicated
Granulocytes	Granulocytes <200/ mm³ with sepsis	Optimum unknown Suggest 10^{10} cells daily × 2	Surgery to drain abscess
Platelets	Platelets <20,000 with petechia	4–8 units (random donor) 2–4 units (HLA type specific)	Avoid drugs affecting platelet function; surgery for specific lesions

Table 7.6
Selected Cancer Chemotherapeutic Agents According to Class of Drug with Incidence and Chronology of Marrow Suppression at Optimal Dose

Drug Class	Examples	Marrow Suppression	Nadir Day	Duration (range)	Comment and Other Dose-limiting Toxicity
I. Alkylating Agents	Nitrogen mustard Melphalan Cylophosphamide (CTX) Chlorambucil	Yes	6–8	4–10 days	Nausea and vomiting Cystitis (CTX)
II. Antimetabolites	Methotrexate 5 Fluorouracil 6 Mercaptopurine Cytosine arabinoside	Yes	6–9	4–12	Stomatitis, renal and hepatic abnormalities
III. Antibiotics	Adriamycin (AD) Actinomycin Bleomycin (BLM)[1]	Yes	10–14	4–7	AD cardiac toxicity[2] BLM pulmonary toxicity[2], dermatitis, fever
IV. Natural Products (Plant Alkaloids)	Vinblastine Vincristine[1]	Yes 0	4–7	3–10	Neurotoxicity
V. Other Compounds	DTIC[1]	Occasional	—	—	Nausea and vomiting, flu syndrome
	Steroids[1]	0	—	—	
	Nitrosoureas BCNU, CCNU, MeCCNU[2] Streptozotocin (STZ)	Yes	20–25	10–40	STZ nephrotoxocity

[1]Non marrow suppressive at therapeutic doses

[2]Maximum cumulative dose limited

Abbreviation DTIC = 5–(3,3–Dimethyl–1–Triazeno) Imidazole–4–Carboxamide

9.0

Summary of Support Therapy Indications

The management of the patient developing anemia, thrombocytopenia, or leukopenia is a common and critically important problem in cancer treatment. The transfusion replacement therapy of the three formed elements is summarized in Table 7.5. Anemia is rarely a clinical problem for two reasons: (1) red blood cell transfusion is universally available; and (2) anemia rarely causes life threatening clinical effects. Leukopenia with secondary infection, however, may lead to severe morbidity and possible mortality. Thrombocytopenia is a more critical clinical problem than anemia but is less acute than leukopenia because platelet transfusions are readily available. Hospitalization is almost never necessary for patients with a decrease in formed elements, with the exception of prolonged neutropenia. However, in the presence of secondary effects like infection and bleeding, hospital monitoring is essential.

121

8 Management of Gastrointestinal Symptoms

1.0

Introduction

Gastrointestinal symptoms are among the most common and the most distressing for patients with cancer. Symptoms may be secondary to inflammatory lesions from the esophagus to the anus and may range from anorexia with nausea and vomiting to diarrhea, constipation, obstipation, and tenesmus. The broad range of gastrointestinal symptomatology may be a direct consequence of the tumor or a secondary effect due to treatment of the tumor. In some instances the gastrointestinal symptoms may be secondary to an unrelated clinical disease such as infection, hepatitis, peptic ulceration, or diverticulitis.

2.0

Differential Diagnosis

Tumors may induce some of the various gastrointestinal symptoms directly by anatomic obstruction in either the large bowel, causing constipation, or in the small bowel or upper intestine, causing vomiting. Another mechanism by which the tumor may cause gastrointestinal symptoms is by tumor secretion of a substance (hormone or other chemical mediator) which secondarily affects the bowel. Carcinoid tumors may result in diarrhea secondary to secretion of serotonin or bradykinin. Secretion of substances by tumors of the islet cells in the pancreas may result in diarrhea (Zollinger-Ellison Syndrome) secondary to gastrin or a vasoactive intestinal polypeptide (VIP). The distinction between the primary and the secondary effects caused by a tumor secretory substance is often dependent upon the identification of the latter in the circulating blood by specific protein assays, and the use of radiographic evaluation of the upper and lower gastrointestinal tract to identify anatomic lesions.

In addition to the gastrointestinal effects which are directly related to the tumor, general metabolic effects may result in any individual gastrointestinal symptoms or in the entire complex of symptoms. Hypercalcemia is one of the common metabolic effects of cancer. In some instances, it is related to the secretion of a PTH-like substance from the tumor while in others to a perturbation of the metabolic calcium balance within the patient's bone. Hypercalcemia may result in a symptom complex which involves anorexia, nausea, vomiting, and constipation. For patients who may be on chemotherapy programs the development of excessive or unusual anorexia, nausea, vomiting, and constipation should suggest the possibility of the development of hypercalcemia.

Another possible metabolic syndrome is related to an indirect effect of the tumor secondary to massive hepatic replacement. In such patients, anorexia and nausea are common symptoms resulting from either a hormone or metabolic imbalance or from the compression of the stomach and small bowel by the enlarged liver.

Chemotherapy drugs are well known causes of gastrointestinal side effects and in particular, nausea and vomiting. The effect of such drugs in the production of the gastrointestinal symptom complex is discussed in detail in the next section.

3.0 **Chemotherapy Induced Effects**
The majority but not all of the known chemotherapeutic agents can cause gastrointestinal effects. In general, such effects are not due to a direct toxic effect on the large bowel, stomach, or small intestine but are related to the stimulation of the vomiting center in the central nervous system. The specific pathophysiologic effects of the drug on the brain are unknown but presumably the drug interacts in some way with the nerve center located in the thalamus. This interaction activates the nerve center and transmits impulses to the higher center of cerebral recognition. At this point, the patient experiences the sensation of nausea. Vomiting occurs with transmission of the sensation input to the parasympathetic nerve trunks which secondarily induces reverse peristalsis from the stomach to the esophagus.

Generally, most drugs induce vomiting within 1 to 4 hours following drug exposure and emesis may persist from 6 to 24 hours. Vomiting almost never extends beyond this period although nausea or anorexia may persist for an additional 3 days. The intensity of the vomiting is variable but may range from nonexistent to continuous vomiting, requiring constant attendance to an available repository. Table 8.1 reviews those chemotherapy drugs which are rarely if ever associated with nausea or vomiting (Column A), drugs only occasionally associated with vomiting but which can lead to vomiting and anorexia (Column B), and drugs uniformly associated with nausea and vomiting (Column C). The most common and well known drugs that cause nausea and vomiting of moderate to severe intensity are the alkylating drugs and the antibiotics. In addition, the drug azacytidine, an antimetabolite employed in acute leukemia, is a profound inducer of emesis. The nitrosourea compounds, experimental drugs used in a wide variety of tumors, are also common offenders.

The chronologic pattern of nausea and vomiting may differ in intensity for different chemotherapy or drug combinations.

124

Table 8.1
Chemotherapy Drugs Associated with Emetic GI Side Effects

A. Rarely associated with GI side effects	B. Commonly associated with GI effects (mild or moderate)[1]	C. Uniformly associated with GI side effects (moderate to severe)
Steroids	Adriamycin	Nitrogen Mustard
Bleomycin	Fluorouracil	Nitrosourea
Vincristine	Procarbazine	DTIC[2]
	Leukeran	Actinomycin D
	L-PAM	Cyclophosphamide
	Methotrexate	
	Vinblastine	

[1]All drugs within this group may produce severe GI effects if the usual doses are exceeded.

[2]Generally less severe with continued exposure (see text).

As previously described, the nausea and vomiting are generally acute but when drugs are employed in multiple drug combinations the vomiting pattern may vary. One particular example is the drug DTIC* which induces severe vomiting episodes during the initial days of treatment. Vomiting disappears by the third day even while the treatment continues, suggesting that the patient's nervous system adapts to the daily drug exposure. This phenomenon is normal for some other drug interactions and is referred to as tachyphylaxis. In intermittent therapy with DTIC, a second course will follow the same pattern of initial vomiting followed by the adaptation mechanism.

Another vomiting pattern is that of the conditioned response. In some patients vomiting occurs the night before chemotherapy in anticipation of therapy. As a consequence of previous treatment the patient develops a Pavlovian response to the thought of receiving the drug and may develop nausea and vomiting just by driving by the hospital. The therapeutic approach to this conditioned response vomiting pattern is different from the treatment of the vomiting induced directly by interaction of the drug with the nerve center. Therapeutic modalities are discussed in a subsequent section.

4.0 **Radiation Induced Gastrointestinal Effects**
Radiation effects on the gastrointestinal tract are related to the dose and volume of radiation delivered as well as to the site of treatment. In addition a generalized syndrome including gastrointestinal effects may develop in patients treated for non-gastrointestinal sites, such as the head or the extremity. Therapy for lung esophageal cancer may result in localized

*DTIC = Dimethyltriazeno imidazole carboxamide, a drug employed in malignant melanoma.

125

esophagitis and pharnyxitis and secondary dysphagia or painful swallowing. Treatment for such effects is generally symptomatic with topical anesthetic agents or analgesics.

Radiation to the abdomen, particularly for ovarian cancer and occasionally for other forms of intra-abdominal malignancy may result in nausea and vomiting, especially for those patients in whom the treatment ports involve exposure of the small bowel. With major nausea and vomiting the therapy must be modulated against the patient's intake. Therapeutic measures other than interruption of therapy are described in the section on antiemetic therapy. Finally, patients treated for gynecologic malignancy or for rectal cancer often develop proctitis and cystitis. These conditions necessitate modification of the radiation dose to decrease the severity of the symptoms. Topical anesthetics for the urinary tract, given by systemic administration, and local measures such as steroids for the rectum may be salutory. A common clinical dilemma in the etiological diagnosis of perirectal symptoms is the distinction between radiation, tumor, and infection related local pathology. In such instances surgical biopsy may be necessary but the biopsy is often negative even in the presence of the tumor.

5.0

Antiemetic Therapy

Vomiting and nausea in cancer patients are most commonly caused by drug therapy. These symptoms occur because the drugs activate the vomiting center of the brain. Therapy must then be directed at inhibiting this center. The major differential diagnosis to be excluded in this clinical setting is gastrointestinal obstruction. The specific therapy for obstruction is surgical intervention.

The gamut of drugs employed in the treatment of nausea and vomiting include antihistamines, anticholinergic drugs, phenothiazines, and combinations of the three. All of the drugs have specific sedative effects on the central nervous system and it is through this mechanism that nausea and vomiting are prevented or ameliorated. These drugs, however, when used alone, are only marginally effective against severe drug induced nausea and vomiting. Even in those instances when nausea and vomiting are only mild or moderate, the antiemetics may be minimally effective. An important aspect of evaluating such drugs is to employ a quantitative scale of vomiting. For example:

Moderate vomiting—defined as not more than two emesis episodes per hour lasting a minimum of 8 hours and associated with constant nausea.

Severe vomiting or *intractable vomiting*—defined as three or more episodes of emesis per hour for a minimum of 8 to 24 hours.

The goal of treatment in patients with either moderate to severe vomiting secondary to chemotherapy is to achieve maximum sedation to the point of somnolence particularly in those patients developing severe vomiting.

126

Standard sedation combined with phenothiazine therapy is successful for a major proportion of patients undergoing drug therapy programs when applied in a chronological sequence in relationship to the treatment. Patients are administered the chemotherapy and at the same time are given barbiturate sedation at a dose of up to 400 mg. If the patient has nausea or vomiting concomitant with the time of treatment, the barbiturate may be given by rectal suppository or alternatively the dose may be divided so that half is given by suppository and half by intramuscular injection. Such a program will induce somnolence within 45 minutes by which time the patient should be home. In this context, it is extremely important to ensure that the patient is accompanied so that he or she will be confident of being transported home safely. For patients who have "conditioned" vomiting as a consequence of prior therapy and in anticipation of chemotherapy, one should prescribe phenothiazine sedation or tranquilizers the night prior to therapy and on the morning of therapy prior to coming to the treatment facility.

The continued use of phenothiazine antiemetics by rectal suppository generally maintains a level of sedation sufficient to prevent vomiting and maintain somnolence. The patient or the family is instructed to administer phenothiazine by suppository or oral tablets at 4 to 6 hour intervals following arrival at home. If the somnolence is profound and sufficient to control or prevent the emesis, additional sedation should not be administered.

Another important aspect of comfort during the period of emesis is to insure that patients eat a full breakfast or meal prior to receiving their therapy. The presence of food in the stomach will diminish the discomfort, should vomiting ensue. The absence of food causes the muscle spasm of retching to contract maximally, and the severity of the pain is proportionally increased.

The major disadvantage of this antiemetic program is the potential danger of pulmonary aspiration should the patient vomit while asleep. In actuality, aspiration has not been observed in many thousands of patients treated on this program. With the level of sedation achieved, patients are usually arousable and may respond to CNS stimulus for emesis with a closed glottis preventing aspiration. A second disadvantage of the barbiturate induced sedation is the hangover which may develop following complete recovery from sedation. In such cases, alternative sedative drugs to induce somnolence should be employed. A summary of the recommended program is outlined in Table 8.2.

One important aspect of the therapeutic approach to emesis is to establish the chronology pattern of emesis response to therapy. The emetic reaction may vary for different drug regimens in different patients. For the first course of treatment, the sedation component of therapy should not be employed. Compazine and other phenothiazines should be used alone. If, for example, nausea and vomiting occur within 1 hour of treat-

Table 8.2
Summary of Antiemesis Program

1. Employ phenothiazines only for initial therapy course to establish pattern of GI response to anti-tumor drugs.

2. Encourage moderate to large meal 1 hour prior to therapy.

3. Tranquilizer and/or phenothiazine the evening before and the morning of chemotherapy.

4. Concomitant with chemotherapy or 1 hour before usual GI episodes administer 300 to 400 mg. barbiturate (may be divided equally into parenteral and suppository).

5. Administer pheothiazines with or without antihistamines every 4 to 6 hours following therapy for 24 hours.

ment then administration of sedation concomitant with tumor treatment is employed. If vomiting develops 4 to 5 hours after therapy then sedation should be administered approximately 1 hour prior to the anticipated or projected commencement of emesis.

5.1

New Methods in Antiemesis Therapy

It is evident that antiemetic therapy, although successful for a substantial proportion of patients, modifies rather than eliminates emesis and is completely unsuccessful for some patients. As a consequence, treatment centers are exploring new approaches both in sedation methodology and in the development of new antiemetic drugs (Table 8.3). One drug of particular interest is the marijuana class of drugs and particularly the active ingredient in marijuana, tetrahydrocannabinol or THC. Marijuana has been known informally to have an antiemetic effect; a secondary but important aspect of the THC effect is appetite stimulation. With the exception of corticosteroids, there are no known drugs that act as appetite inducers and therefore are helpful in the anorexia symptom. This information led to a double blind study of THC. In the study, THC was compared to a placebo in a tumor chemotherapeutic program which induced major and consistent nausea and vomiting. THC was superior to placebo and proved to be more effective than phenothiazines in standard dosage in a later study. Patients treated with THC achieve a mild to moderate euphoria which appears to be an essential aspect of the antiemetic therapeutic effect. At present THC is in the process of development and is under FDA regulations so it is not readily available.

Another group of drugs which are well known as antiemetic

Table 8.3
Antiemetic Methods in Association with Cancer Chemotherapy

	Method	Types	Schedule
Commonly Used Drugs	Prochlorperazine (Compazine)	Parenteral	Q4–6h
		Suppository	Q4–6h
		Oral	
	Trimethobenzamide (Tigan)	Parenteral	Q4–6h
		Suppository	
		Oral	
	Antihistamines	Oral (Liquid & Capsule)	Q4–6h
	Barbiturates	Parenteral Oral	See *Antiemetic Therapy*
New Drugs	THC (Tetrahydro-cannabinol)	Oral	See *Antiemetic Therapy*
New Concepts	Mind Control	Self Hypnosis	See *Antiemetic Therapy*

agents and also as amnesic agents is the anticholinergic group and in particular scopolamine. Such drugs have major adverse effects due to central nervous system disequilibrium syndromes but potential development of analogues or improvements in methods of administration may result in practical application of these new drugs.

In addition to drug therapy an innovative approach to emesis control is meditation and mind control. The objective of this approach is to develop an inner sense of well-being and peace and to positively reinforce an avoidance of emesis. Whether employing yoga methods or meditation, all such self-control thought experiences may be regarded as forms of self-hypnosis. It is not well known that self-hypnosis has been used in an effort to control specific symptomatology induced by chemotherapy. Hypnosis has been successful in treating patients with habits, such as smoking or obesity due to overeating. It is expected that hypnosis can also be used to control the conditioned reflex-type vomiting induced by a Pavlovian response. In preliminary studies of the effect of hypnosis on such patients, mind control effectively eliminated the nausea and vomiting induced by pretreatment anxiety. It also ameliorated the nausea and vomiting that followed chemotherapy treatment. Hypnosis demands acceptance and commitment from the patient; the hypnotherapist requires four to five training sessions in order to develop the technique. Some patients may be more susceptible and receptive to hypnosis as a methodologic approach. Those patients who are artistic, sensitive to

129

aesthetics, interested in the supernatural or in astrology lend themselves readily to hypnotic management. Organically depressed patients or patients with extreme egos or concepts of independence are generally not receptive to mind control approaches.

6.0   **Bowel Function Therapy**

This section will focus on management of diarrhea, constipation, and the unusual symptom of tenesmus. Such symptomatology is almost invariably related to a pathologic process in the large bowel as opposed to the small bowel where nausea and vomiting are more common clinical manifestations. Diarrhea may be one of the most distressing symptoms the patient may have to deal with and if an organic cause is not established that is related to some hormonal substance secreted by the tumor, then bowel control should be achieved by employing narcotics. Two important causative factors must be ruled out, however. Fecal impaction with diarrhea stool developing around a local obstruction should be identified and treated by disimpaction. A second potential cause of diarrhea is the stool incontinence associated with spinal cord compression which is generally diagnosable by neurologic examination. If after ruling out these two causes diarrhea persists, then treatment with anticholinergic drugs in sequence and opium derivatives in the form of paregoric will control bowel function. The most critical feature of this treatment is to employ adequate dose and frequency of the opiate which may at times necessitate administration of the drug at 2 hour intervals.

Constipation is the bane of the elderly and is a common accompaniment to old age related to sluggish peristalsis. Elderly patients who are most prone to malignancy commonly develop irregular stools. Such irregularity may be compounded by the use of narcotics for pain control which results in additional constipatory problems. These patients should be put on a regular bowel regimen. This regimen includes daily laxatives and dietary supplements which are designed to promote bowel motility. The addition of stool softeners to create more liquid movements may also be recommended. Such a regimen is particularly important for patients receiving narcotic therapy. Again, the various laxative preparations which may range from gentle irritants to major propellants should be employed before turning to the use of enemas. Constipation may develop into obstipation which may be defined as the complete inability to develop bowel motility in the absence of anatomic obstruction. Some chemotherapeutic drugs, particularly the periwinkle alkaloids, may create such a circumstance by causing a parasympathetic neuroplegia and an adynamic ileus. The clinical picture may mimic bowel obstruction and can result in the need for surgery. Patients receiving such drugs should be monitored closely and cathartics should be employed prophylactically, particularly in elderly patients.

Tenesmus may be defined as the constant sensation of rectal fullness and the need to evacuate the bowels. This sensation, which is extremely distressing to patients, is a reflection of the presence of a mass lesion in the rectum generally compressing the sacral plexus but also distending the rectal pouch locally. The resulting sensation is the urge to defecate. Such a sensation may or may not be associated with actual pain. The management of such a problem is contingent upon decreasing the local expansive pressure by a direct anti-tumor program such as radiation. Bypass surgical procedures with a colostomy may be required for definitive management although such surgery may not relieve the sensation of discomfort.

7.0   **Ostomy Management**

One of the greatest fears of the cancer patient is the fear of having an ostomy. This fear is primarily a reflection of the distaste for scatological functions and a transposition of such functions to a more obvious anatomical site. A second fear is that the odor and management of the bowel function will be beyond control and will interfere with the patient's normal life style. Actually, patients with ostomies related to cancer almost invariably have a colostomy as opposed to an ileostomy and a major feature of colostomies is that bowel control can be achieved and often without the need to use a cellophane receptacle.

Colostomies may be a temporizing measure which allows the large bowel to be decompressed. When the local inflammation and edema of obstruction subside re-anastomosis of the bowel is possible. For patients with rectal cancer, however, a permanent colostomy may be necessitated by the need for removal of the entire distal portion of the colon. Thus, permanent colostomies are done for patients with low rectal and recto-sigmoid lesions and are generally placed in the left lower quadrant. In general, colostomies should be placed in a position that is away from the belt line and away from the site of body flexion.

A review of colostomy management and the life styles of patients in relationship to the ostomy is available from the United Ostomy Association and should be provided to all patients requiring ostomy for instruction in care and reassurance about life style.

8.0   **Ascites Management**

Ascites or the development of intra-abdominal fluid is a common complication of malignancy, particularly in patients with ovarian cancer. It also develops in patients with colon cancer, pancreatic cancer, and primary or metastatic tumors of the liver. There are two pathophysiologic mechanisms in the production of ascites: (1) increased pressure in the portal venous drainage system with secondary transudation of fluid or (2) exudation of a protein rich fluid directly from diffuse peritoneal implants of tumor. Although the former mechanism is uncommon in ovarian cancer, it occurs frequently with intrahepatic tumors.

The major problems in ascites management are pain with

distention and malnutrition caused by the protein deprivation resulting from pooling in the ascitic fluid. Fluid compression of various organs within the abdomen may result in decreased appetite and abnormal bowel function. There are three therapeutic approaches to ascites. First, the tumor may be specifically controlled, an approach which commonly works for ovarian cancer but which is not successful for tumors which are less responsive to therapy. Second, intra-cavitary cytotoxic drugs or radioactive substances may be used which may cause a local inflammatory reaction and close off the peritoneal space. This approach is fraught with major secondary effects and is not generally effective therapeutically. Third, a shunt may be created from the abdominal cavity into the venous access system by a catheter from the peritoneal surface, under the skin to the jugular vein. The patients most suitable for the latter approach are those requiring paracentesis of 4 or more liters of fluid weekly. This method of management is investigational but has major advantages for patient comfort and nutrition.

9.0   **Jaundice Management**

Jaundice may result from direct damage to the liver caused by the tumor or from obstruction of the biliary tree draining the liver. The cutaneous pigmentation is directly related to the deposition of bilirubin in the skin as a consequence of increased circulating blood levels. Bilirubin per se does not have an adverse effect on the host systemically and does not lead to any specific symptomatology. However, the jaundice does have a major cosmetic effect and is often disturbing to the patient. In patients with jaundice, particularly if associated with obstruction of the biliary tree, the accumulation of bile salts in the circulation will lead to cutaneous deposition and may lead to pruritis. The symptom complex of pruritis-excoriation may be severe and may be manageable only by a surgical bypass of the biliary tree to ensure that the bile salts normally secreted by the liver do flow into the gastrointestinal tract and are not retained. Local obstruction may also be relieved by radiation to the porta hepatis. The diagnostic distinction between intrahepatic disease and biliary obstructive disease in jaundiced patients is made on the basis of the presence of clinical pruritis. Treatment, if surgical bypass procedures are not possible, may be symptomatic (using antipruritic drugs) or alternatively the use of cholestyramine to bind bile salts within the bowel lumen may be effective. The life expectancy of patients developing jaundice secondary to malignancy may be so limited that major surgical intervention is not indicated and simplistic palliation of symptoms should be employed.

10.0   **Anorexia Management**

Anorexia is one of the major symptoms of cancer. A depleted nutritional state secondary to a decreased appetite may perpetuate cancer growth by depriving the host of the necessary nutrients to maintain host defenses against the tumor. This vicious cycle is difficult to interrupt therapeutically but hyperalimentation, both intravenous and enteral, may be successful in compensating for anorexia while not directly reversing it. The chapter on management of nutrition reviews the biochemical disruption created by cancer cachexia and malnutrition as well as the specific therapies available. Unfortunately, treatment for anorexia is particularly difficult because of the complexity of the components of food appeal which include taste, smell, and central gustatory appreciation—all of which are disrupted in the cancer patient.

11.0

Accessory Organ Malfunction

The accessory organs of the gastrointestinal tract include the exocrine glands such as the salivary gland and the pancreas. These organs are commonly affected by therapy and rarely by the tumor. Salivary gland dysfunction results primarily as a consequence of radiotherapy which may lead to xerostomia (dry mouth) and secondarily decreased taste capability. This effect compounds the problem of anorexia and food intake and requires supplemental liquids to soften and moisturize foods. In some patients artificial saliva may be used but is rarely helpful because of the frequency of application that is necessary. Occasionally, patients require the constant availability of drinking material to be able to swallow the normal mucous secretions of the nasopharynx.

Pancreas dysfunction as a consequence of tumor invasion of the organ can result in depletion of exocrine or digestive enzyme function, as well as the endocrine function, particularly insulin. The use of enzyme supplements, such as viokase, to replace endogenous secretion will invariably control the diarrhea associated with pancreatic insufficiency. The dose and schedule of enzyme replacement are important to ensure adequate digestion. Insulin replacement is rarely necessary unless a total pancreatectomy is performed and in such patients the insulin requirement is modest.

12.0

Summary

Gastrointestinal symptoms in cancer patients are a reflection both of the tumor and of the therapy, and represent some of the most distressing symptoms for the patient to endure. Although there are a great number of therapeutic modalities employed for the control of such symptoms, the success of therapy is to a large extent dependent upon recognizing the specific cause. Surgery is required for patients with local obstruction, resulting in constipation, or with local obstruction, resulting in nausea and vomiting. For patients with metabolic or drug induced causes of emesis or diarrhea, control with standard forms of sedation is only partially effective. The most important ingredient in the management of gastrointestinal symptoms is the application of prophylactic intervention.

9 Management of Oral Ulceration

1.0

Introduction

The oral cavity is a common site of the origin of malignancy. It is also commonly affected by secondary infection and therapeutic modalities, such as radiation and chemotherapy. The symptom complex resulting from oral pathology includes pain and the inability to swallow. The secondary effects of compromised deglutition or swallowing are a decrease in food intake and an impoverished nutritional state which contributes to the inability of the local ulcerations to heal. The management of mouth care is therefore an extremely important aspect of cancer care, not only for patients with cancer of the head and neck, but also for patients on therapy programs which may cause secondary oral irritation. This chapter will focus on the diagnosis and treatment of oral ulcerations primarily induced by chemotherapy and on prophylaxis programs for oral hygiene to prevent or minimize local mucositis.

Stomatitis and mucositis are synonymous for the more descriptive term of oral ulceration. When a person is afflicted with an ulceration, an inflammatory lesion is present within the oral cavity, often in association with depression of the surface epithelium. With the exception of the anatomical distribution, oral ulcerations are generally local manifestations of a variety of pathologic insults to the oral mucosa and are not diagnostic or characteristic for a specific disease.

2.0

Predisposing Factors to Oral Ulceration

There are a number of predisposing drug and host tissue factors which may make the patient susceptible to the development of oral ulceration. The foremost predisposing factor is the patient's natural oral hygiene. Patients with dental cavities or underlying inflammatory gingival disease or chronic low grade mouth infection are especially prone to secondary stomatitis induced by drugs. Such oral hygiene is often associated with chronic smoking and alcohol ingestion which in turn may generally be correlated with the secondary development of head and neck cancers.

A major inducer of oral ulceration or oral inflammation is radiotherapy delivered to the oral cavity. Radiation not only causes acute stomatitis by affecting the mucosa, but may bring about salivary gland dysfunction. The production of saliva may decrease, thereby altering the natural acid base balance and general oral milieu. Such effects by radiation are often irreversible and in the presence of a secondary insult with chemotherapy, the oral lining may be predisposed or sensitized to an intensified stomatitic response. Therefore, methotrexate or other drugs which commonly produce stomatitis in patients who have previously received radiotherapy may also produce oral ulceration when a lesser dose of the drug is given. In addition, some drugs such as adriamycin which are radiosensitizing may cause a recall of previous radiation damage and this effect in association with the usual ulcerogenic effect of adriamycin may result in major oral pathology.

3.0   **Chemotherapy and Oral Ulceration**

Anti-tumor drugs affect the normal mucosal lining of the oral cavity by inducing an inflammatory reaction which is a common side effect of some anti-tumor agents. Antimetabolites are the most common agents that cause secondary oral ulceration. In general, alkylating agents and plant alkaloids do not result in oral ulceration. This unique difference between antimetabolites and other drug classes may be due to the fact that antimetabolites primarily attack cells undergoing rapid turnover. The cells lining the mucosal surface are rapidly replicating cells. Methotrexate profoundly affects the oral tissue. One can thus argue that methotrexate could be used to treat squamous cell carcinoma tumors in the head and neck area that evolve from the oral mucosa.

The mechanism for the production of oral ulceration is by the inhibition of the basal layers of the mucosa preventing replication and replacement of the most superficial cell structure. Thus, cells dying off in the most superficial layer in a senescent state are not replaced by the deeper cell layers which would normally maintain integrity of the oral mucosa. It is important to understand that the lining mucosa of the entire gastrointestinal tract, including the rectal and peri-anal area, the vaginal orifice, and the glans penis, are lined by similar mucosal tissue and that ulcerations developing in the mouth are often accompanied by ulcerations in the genitalia and the gastrointestinal tract. The general pattern of chronology in the development of ulcerations is conditioned by the normal cell reduplication and replacement system in the oral cavity. Ulcerations generally appear 5 to 7 days following drug exposure and may persist for as long as 4 to 10 days. The severity of the mucosal effect may vary from paresthesias or dysesthesias, without overt interruptions of the mucosal integrity, to total

136

surface denudation with confluent ulceration of the entire oral cavity including the palate, naso-pharnyx, peri-tonsillar, and buccal and gingival mucosa. Generally, ulcerations secondary to drug therapy are discrete and most commonly involve the following areas: the buccal mucosa in apposition to the molar teeth, the subglottic area at the sites of entrance of the sub-mandibular salivary glands, and the peri-tonsillar and posterior palate area, including the uvula. Ulcerations generally measure 2 to 5 millimeters and are superficial.

The drugs most commonly associated with oral ulceration are 5-Fluorouracil and Methotrexate (antimetabolites), and Adriamycin and Actinomycin D (antibiotics). The devastating effect of these toxic drugs is severe oral pain. In addition the patient may restrict nutritional intake, occasionally requiring tube or intravenous feeding to maintain adequate nutrition and hydration. The ulcerations are always reversible, and palliative measures should be employed while awaiting reconstitution of the mucosal integrity. The three ulcerogenic drugs most commonly applied are described in detail with reference to oral ulceration.

1. *Adriamycin* is an anthracycline antibiotic which causes oral ulceration when employed in doses over 60 mg/M²/ treatment. It has been demonstrated that the ulcerations occur less frequently if the drug is administered as a single dose as opposed to a 3 day divided dose course. The ulcerations are commonly associated with pharangeal discomfort in the absence of visible ulcerations, and the localization of the discomfort may be in the retro-oral area rather than in the buccal mucosa and tongue.

2. *5-Fluorouracil* is an antipyrimidine which has as a dose limiting side effect mucosal irritation, ulcerations, and generalized systemic gastrointestinal effects. The ulcerations observed with 5-Fluorouracil are commonly seen in the anterior portions of the mouth and over the gingival surfaces as well as the palate. The ulcerations are generally superficial and recede promptly with drug withdrawal. The intensity of the effect on the mucosal lining is directly proportional to the dose of drug administered and is generally of mild to moderate intensity with the usual therapeutic drug dosages. When the drug is administered as a continuous infusion the dose limiting toxicity is still oral ulceration but at a much higher daily dose.

3. *Methotrexate* is a folic acid antagonist which affects the oral mucosa in much the same manner as the antipyrimidines, particularly with regard to the distribution of oral ulcerations. Recently, high doses of Methotrexate have been employed in conjunction with the administration of the antidote (folic acid) which protects the normal mucosal cells as well as the marrow cells from the cy-

137

totoxic effect of the drug. The development of mucositis, however, may occur despite the systemic administration of the antidote. Recent investigations have explored the use of topical folic acid or topical thymidine (administered in a paste vehicle) for preventing ulcers. These techniques are particularly efficacious for patients who develop ulcerations on the vermillion border of the lip. These ulcerations are extremely uncomfortable and are often associated with major secondary inflammation.

4.0

Other Causes of Oral Ulceration

The differential diagnosis of intra-oral ulceration is often complicated by two factors: 1) the lack of a pathognomonic feature to distinguish one form of oral ulceration from another, and 2) the development of such ulcerations in a multi-factorial circumstance. Secondary infection or superinfection with opportunistic organisms such as monilia may occur. Treatment for each of the different types of stomatitis may be different and specific. For example, in a patient with acute leukemia who is being treated with ulcerogenic drugs, the development of stomatitis in association with neutropenia and active leukemia may make it impossible to distinguish drug-induced ulceration from secondary superinfection or leukemia infiltration, all of which may in fact be present simultaneously. The various types of mucositis or inflammatory stomatitis are described in Table 9.1. A pictorial review of some of the patterns of stomatitis is described in Figure 9.1.

Table 9.1
Differential Diagnosis of Oral Ulceration

Herpes simplex	Punctate lesions with/without vesicles, inclusion bodies intracellular
Apthous stomatitis	Large lesions, confluent, involves palate and buccal mucosa
Monilia stomatitis	White exudate over ulcerative base, involves palate, nasopharynx, and buccal mucosa, typical yeast forms on wet preparation. Confluent denudation with yellow, pustular exudate involves gingiva and peritonsillar area
Infiltrative (Neoplastic, usually leukemia)	Indurated borders around an ulcerating base, biopsy demonstrates leukemic cells

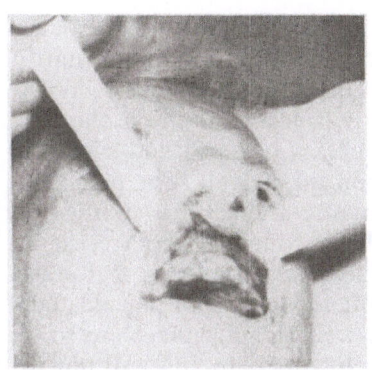

Figure 9.1
Hemorrhagic ulcerative lesions involve the vermillion border of the lip and buccal mucosa, but spare the gingiva. These lesions are typical of toxic reaction to the antimetabolites (FU and Methotrexate) and their localization and appearance is distinctive from that of monilia (thrush) infection.

Apthous stomatitis is a form of oral ulceration characterized by large patches of ulceration over the buccal, gingival, and palatal surfaces. The etiology of this form of stomatitis is unknown, but it does not occur in greater frequency in patients with malignancy. Therefore, it should not be considered as a usual cause of stomatitis in patients undergoing treatment of neoplastic disease.

Herpes simplex stomatitis is caused by a DNA virus and such infections are commonly associated with vesicles which may distinguish them from other oral ulceration disease. The vesicles may extend over the lips and higher tonsil areas as well as to other sites in the oral cavity. Anti-viral agents have been employed, although unsuccessfully, in the treatment of such lesions. Specifically, iododeoxyuridine (IUDR) and cytosine arabinoside are two agents which have been used topically and systemically, respectively, with marginal success. More recently, adenine arabinoside, a drug which is specifically effective in herpes type viral infections, has been employed with a greater degree of success, particularly in herpes zoster infections, in reducing morbidity of infection when employed early in the course of disease.

Monilia stomatitis is caused by a fungus with a predilection for mucosal surfaces. This infection is common in children (known as thrush) as well as being a frequent cause of vaginitis in adults. The organism is ubiquitous and is a common surface contaminant in symbiosis with humans. In the event that natural defense mechanisms are depressed, as in cancer, an invasive and/or systemic infection with candida may develop. Monilia stomatitis, the major manifestation of monilial infection, commonly affects the esophagus as well as the oral mucosa. Patients who have a burning sensation in their esophagus and show no signs of oral moniliasis should be examined for esophageal moniliasis. In oral moniliasis, the lesions in the mouth are commonly covered by a white coating representing the colonies of the organism and are millet seed in distribution. Under the white secretion is an ulcerative base which is discrete, erythematous, and tender. The treatment of monilia infection of the oral cavity is specific and therefore the diagnosis should always be sought. The diagnosis is made by identification of the monilia yeast forms in wet slide preparations prepared directly from the lesion.

The intra-oral tumor is not included in this list of the causes of oral ulceration. Metastases from sites outside the oral cavity are uncommon. The exception to this rule is the leukemia group of diseases which are frequently associated with mucosal invasion throughout the GI tract including the oral mucosa. If indeed leukemic infiltration is established by a biopsy or touch preps, then local radiation therapy can control the local invasion. Primary tumors arising within the oral cavity by and large demonstrate an exophytic lesion or an irregular infiltrative area on the mucosal surface. Such lesions are commonly painful and result in a rigidity of the underlying surface.

Treatment of Stomatitis

Unfortunately, the treatment of oral ulcerations is primarily symptomatic and generally has limited effectiveness. Withdrawal of cytotoxic chemotherapy is essential and is the first step in aborting the mucosal ulceration. Local and systemic measures for pain relief are important as well. The pain may be severe enough to require narcotic analgesia systemically. The use of such drugs should be encouraged as a necessary although temporary measure. Generally, patients feel most comfortable eating fluids and soft foods of cool or cold temperature. Hot and spicy food as well as acid juices are to be avoided. Ice is often helpful in relieving pain and will not interfere with healing.

Local methods of analgesia are only temporary but can provide major relief. Viscous xylocaine preparations have been employed by and large without success. The local application, however, of an anesthetic paste to individual ulcerations by a cotton swab will relieve pain immediately and the pain relief may be sustained for 30 to 40 minutes. Unfortunately, the paste therapy is accompanied by a local anesthetic feeling inhibiting the ability to taste food and creating local dysesthesia. Nonetheless, such pastes are the most successful pain control measures to be employed.

Secondary bacterial and superinfection should be specifically managed by appropriate antibiotic or antifungal agents. Tetracycline may be diluted in water and used as a gargle for patients with secondary infection and may also be superficially applied to the mucosal surface in patients with culture proven pathogenic infection. With secondary fungal infection, mycostatin solution should be administered by a dropper; 2 cc of the solution should be placed in the mouth every 4 hours and swallowed slowly. Gentian violet, an antiseptic solution applied by swabs to the entire buccal mucosa, is also employed in conjunction with mycostatin. The latter should be applied frequently to maintain coverage of all lesions.

Chemotherapy induced ulcerations are transient. Palliative or supportive therapy through the period of maximum discomfort is temporary. The application of such therapy to other mucosal sites outside the oral cavity, such as the vagina or peri-rectal area, is often not as successful as in the oral cavity because of the difficulty in local application. Nonetheless, such measures should be applied to all mucosal sites to control the local irritation.

One of the most critical factors in the treatment of stomatitis is to anticipate the development of such abnormalities, particularly in patients who may be predisposed to develop ulcerations, such as those with poor oral hygiene. In these instances, prophylactic mouth care, particularly in the form of gingivectomy or vigorous oral cleansing prior to initiation of treatment, is essential. Dental extractions, antibiotic therapy, and periodic hydrogen peroxide mouthwash may be helpful in minimizing local inflammatory lesions which would accentuate the effect of chemotherapy induced ulcerations.

In the presence of ulcerations, mouth care must be continued and soft plastic sponge toothettes can be employed for cleansing as well as debridement of ulcers. One should use a non-astringent saline wash to clean the teeth as well as the oral mucosa.

10 Management of Pain Syndromes

1.0

Introduction

Pain is the overriding fear of the cancer patient. Frequently, those patients who accept the diagnosis of cancer most easily will ask first and foremost: "Doctor, will it be painful?" Actually, in the course of the disease, pain is usually controllable and does not interfere with the patient's life style. Although patients can be reassured that pain as a terminal event is not common, there are, nonetheless, a number of clinical situations which acutely require pain management and a proportion of patients for whom pain management is the crucial issue in the clinical course.

This chapter will review the pathophysiology of pain and focus on a number of clinical examples of common and rare pain syndromes developing as a consequence of cancer metastases.

2.0

Theories of Pain Perception

Over the past century three theories of pain perception have evolved. The "specificity" theory suggests that specific receptors for sensations of cold and warmth, touch, and pain are present in the peripheral integument. The "pattern" theory, developed at about the same time, suggests that pain is perceived as a consequence of the intensity of the stimulus in combination with multiple stimuli. Pain therefore results from an accumulation of many factors. Finally, the "gate" theory, which is the most widely held at this time, proposes that there are two peripheral sensory afferents: one is large and the other is made up of C fibers or slow pain response fibers. In addition, a central cerebral modulating system emanates down through the spinal cord to the level of entrance of the large and C fibers. The central modulating area is the area of the nervous system upon which most analgesic drugs administered systemically act.

These simplistic explanations of the pathophysiologic mechanisms of pain do not take into account the major ancillary

143

issue of the psychogenic component of pain: patients may accentuate their pain by focusing their attention on the effect of pain on their life style. Generally, anxiety and anticipation lower the pain threshold, and psychologic depression reinforces the intensity of pain. Therefore, as we will see in a subsequent section, in addition to analgesics, drugs affecting mood are important additions to the drug armamentarium in treating pain syndromes.

3.0 **Clinical Aspects of Pain**

The diagnosis of pain etiology is dependent upon an analysis of the six clinical features of all pain. First, it is important to distinguish acute from chronic pain and to determine if the pain is directly related to cancer or is secondary to a residual structural abnormality. For example, a persistent structural defect in the lumbar spine, which remains after radiation has effectively eliminated the tumor, may cause continuing pain. In this situation, continued radiation therapy is not employed but orthopedic procedures to maximize local support are effective.

Intensity or severity of pain may be quantified relative to the incapacity the patient experiences. For example, pain requiring cessation of all activity or lying motionless in bed is a quantitative measure and is more severe than pain that restricts activities marginally. Another scale that the interviewer may use is to ask patients to rate their pain on an arbitrary scale of one to four or one to ten, so that patients can relate the intensity of the pain to their own experience.

Pattern or the evolutionary character of pain is an important clinical aspect of pain. Pain characterized by an intermittent intensity crescendo pattern is characteristic of colic or cramping pain and suggests an obstruction to a tubular visceral organ, such as the GI tract or the biliary tree. Other pattern characteristics are intermittent versus constant pain or nocturnal versus morning pain. These features may be helpful in establishing the pain etiology.

Duration of the pain is an important quality and contributes to diagnostic information. Constant unrelenting pain suggests a mechanism of contiguous pressure of the tumor directly on nerves. Intermittent, sudden, and short acting pain implies that the tumor is stimulating nerves secondary to a precipitating factor possibly related to position, food, or other factors.

Response to therapy is important in quantifying the intensity of the pain and is also helpful in determining the direction of further explorations in diagnosis. Thus, pain that is responsive to salicylates may be considered to be inflammatory in etiology and is quantitatively of less importance. On the other hand, pain unresponsive to narcotic analgesics suggests either permanent structural damage or progressive tumor growth.

Site of the pain is of obvious importance in the diagnosis as well as in therapy for pain. Referred pain may confuse the specific pain localization. For example, pain which is felt by the patient to be localized to the hip or flank may actually derive from a bone lesion in the shaft of the femur. When evaluating patients with poorly defined or possibly referred pain, a general evaluation is important. Another common type of referred pain is shoulder pain which results from either an infarcted and enlarged spleen or from distention of the liver secondary to hepatic metastases. Diaphragmatic irritation and stimulation of phrenic nerve afferent pathways to the shoulder is the presumed mechanism for shoulder referral localization. Non-referred pain, however, is more common. In general, pain serves as a guide to localization of therapy. For example, if there are multiple sites of bone abnormality found by x-ray or scan, the pain will indicate where treatment should be specifically applied.

Identification of precipitating factors is especially important in attempting to establish the etiology of pain. Pain may be precipitated by specific positions. For example, the back pain associated with pancreatic cancer may become more prominent when bending over. Food or alcohol may precipitate pain in the area where the tumor is localized. This is characteristic of Hodgkin's Disease, although it is an uncommon phenomenon.

The patient experiencing pain should be confronted with each of the above questions in order to characterize maximally the features of the pain which may lend themselves in turn to a specific diagnosis and means of management.

4.0
Types of Cancer Related Pain
As indicated previously, the pain syndrome caused by cancer may be catagorized simplistically into acute pain syndromes and chronic pain syndromes. A third category of unrelated or incidental pain syndromes secondary to ancillary or co-morbid disease may also be considered.

Within the *acute* pain syndrome category, six clinical examples are described which represent the common acute pain syndromes of metastatic cancer.

Bone Pain. Bone pain is the most frequent form of acute pain and may or may not be associated with pathologic fracture or structural defect. The pain may be fluctuating in intensity and frequency related to climatic conditions, much like arthritis, and may remit spontaneously for some period of time. Often the pain is caused not by disruption of the integrity of the bone cortex, but rather from distention within the marrow cavity. In this setting, pain may be exquisite and no radiographic changes may be observed. Localization by physical examination, however, pinpoints the lesions to be within the bone and a bone marrow biopsy will often reveal the diagnosis. Patients with localized bone pain are by and large managed with local radiotherapy which relieves the pain in the majority of patients, even if major tumor regresssion and bone healing are not observed. The implication is that the minor decrease in tumor size is sufficient to relieve the pain but not sufficient to allow remodeling and reconstruction of the bone. In some circumstances the tumor may be completely eradicated by the tumor-specific therapy but because of disruption of the cortex and collapse of the bone, the structural defect remains and may be potentially compressing local neuro structures resulting in pain. In such patients back supports or splints may be necessary to augment the structural integrity of the bone to maintain a pain-free state.

Spinal Cord Compression. This clinical syndrome is often heralded by back or radicular pain which follows the nerve route distribution within the dermatome. In all patients with back pain, the potential for spinal cord compression must be considered. Appropriate diagnostic and therapeutic steps are initiated on an emergency basis because of the potential for rapid progression of the clinical picture with impingement of the tumor directly upon the spinal cord with paralysis and irretrievable loss of bowel and bladder function. Radiotherapy is often used for pain secondary to spinal cord compression. In the early pre-morbid stages, however, decompression laminectomy is an essential component to insure tumor control. An interesting physical finding is the presence of localized tenderness over the vertebral body under which the tumor lies within the dural space. Patients often comment that flexion of the body results in accentuation of the pain and occasionally in a specific radiation of pain down the body, called the Lhermitte's sign.

Serous Effusions (excluding ascites). The intra-thoracic pleural effusions, either pericardial or pleural, generally cause pain only indirectly as a consequence of a secondary inflammatory reaction. The pain syndrome is actually characteristic of the benign inflammatory effusions that may occur in these potential spaces, characterized by a pleuritic component or by accentuation with position, particularly pericardial effusions and forward positions of body flexion. Occasionally, the serositas associated pain may be more intense secondary to tumor implantation along the intercostal nerves which are in direct continuity with the pleural surface. Pain management in these instances is directly related to and facilitated by control of the tumor. Pleural effusions are commonly caused by and are associated with ovarian or breast cancer. Since these tumors are exquisitely responsive to a variety of forms of treatment with cytotoxic drugs, successful pain control is often achieved for a period of time. For pain secondary to pericardial effusion, however, radiotherapy is often necessary for pain and tumor control.

Liver Distention. A common site of metastases for a multiplicity of tumors is the liver. Pain as a secondary consequence of liver metastases is unusual in the absence of massive hepatomegaly with distention of Glisson's capsule which envelops the liver. Such pain is often referred to the shoulder as a consequence of diaphragmatic irritation and may be interpreted as bursitis or arthritis. Specific management, again, is dependent upon control of the tumor and local radiotherapy or hepatic infusion therapy with cytotoxic drugs may be helpful in a substantial proportion of patients.

146

Ascitic Distention. Ascites is another common complication for a multiplicity of tumors, including ovarian cancer. Massive abdominal distention with stretching of the cutaneous surface causes mild to moderate pain and discomfort. Generally, management is accomplished by simple fluid removal which, however, may require weekly paracentesis (see gastrointestinal effects of malignancy for a discussion of alternative management of ascites).

Headache. Headache is a common symptom in patients with cerebral metastases. Any patient with a headache which is not localized or which is focused in the apex or vertex of the scalp, and which is resistant to mild analgesics, should be evaluated appropriately for CNS metastases. It is exceptional for headaches to occur in an isolated fashion without neurologic symptoms such as blurred vision, papilledema, or focal peripheral neurologic abnormalities. Therapy is focused on the use of steroids to reduce the peri-tumor edema, or the application of radiation therapy or surgery to decrease the tumor size directly.

Chronic pain syndromes induced by cancer are more recalcitrant to direct tumor specific therapy modalities than the acute pain syndromes. For this reason, chronic pain syndromes require analgesic drugs or neurosurgery. Four specific syndromes of chronic pain have been selected for discussion. They are representative of the important types of syndromes and all occur frequently.

Bone Pain. Structural defects may bring about a chronic pain syndrome. Although this pain is almost always associated with an established tumor lesion, it is difficult to localize. Structural supports are employed for treatment, particularly for compression of the vertebrae secondary to pathologic or healed fractures.

Pancreas Pain. The retroperitoneal localization of the pancreas commonly results in extension of the tumor into the paraspinal musculature and neuroplexies. As a consequence, pain is a common accompaniment of pancreatic cancer. Although radiotherapy is occasionally palliative for this condition, the unresponsiveness of the tumor results in rapid regrowth of the tumor and return of the pain. Narcotic analgesics are generally necessary, while neurosurgery is rarely necessary because patients die so rapidly.

Rectal Pain. Carcinoma of the rectum is often associated with perineal sacral plexus invasion and pelvic pain is common particularly in the presence of a large volume of tumor. Again, radiotherapy is often employed but is only temporizing. Neurosurgical procedures or chronic analgesia are frequently necessary to control the pain. Colon bypass procedures (a colostomy) are useful to decompress obstruction, but do not alleviate the pain which is secondary to tumor compression or invasion of the nerves.

Extremity Pain. Extremity pain may derive from proximal obstruction such as when a superior sulcus tumor compresses the brachial plexus in lung cancer or when breast cancer metastatic to the brachial plexus area occurs in the supraclavicular fossa. The latter circumstance is also often associated with major lymphedema of the extremity. Radiation therapy is primarily used to manage these forms of pain due to proximal tumor extension. Another form of extremity pain secondary to cancer is associated with mass lesions in the subcutaneous soft tissue of the distal extremity. This condition occurs particularly with soft tissue sarcomas and malignant melanomas. In these circumstances, the tumor may reach such a large size as to necessitate amputation to control pain. Balancing limb function against the severity of symptoms secondary to the tumor is a difficult problem to resolve.

Subcutaneous metastases may be managed by local cryosurgery or electrocoagulation. Figure 10.1a depicts a patient with a metastatic sarcoma of the buttock, a condition accompanied by severe pain and ulceration. With local electrocautery, the lesion was excavated (Figure 10.1b) and healed promptly without residual discomfort (Figure 10.1c & d). Another example is the patient with cutaneous melanoma similarly affected by pain and ulceration with an exophytic lesion (Figure 10.2). Palliation by electrocauterization minimizes the operative morbidity with excellent local tumor control.

Atypical Pain Syndromes

5.0

There are a number of rare or atypical pain syndromes that may be specifically related to or associated with cancer but which may not be a direct consequence of the tumor. For ex-

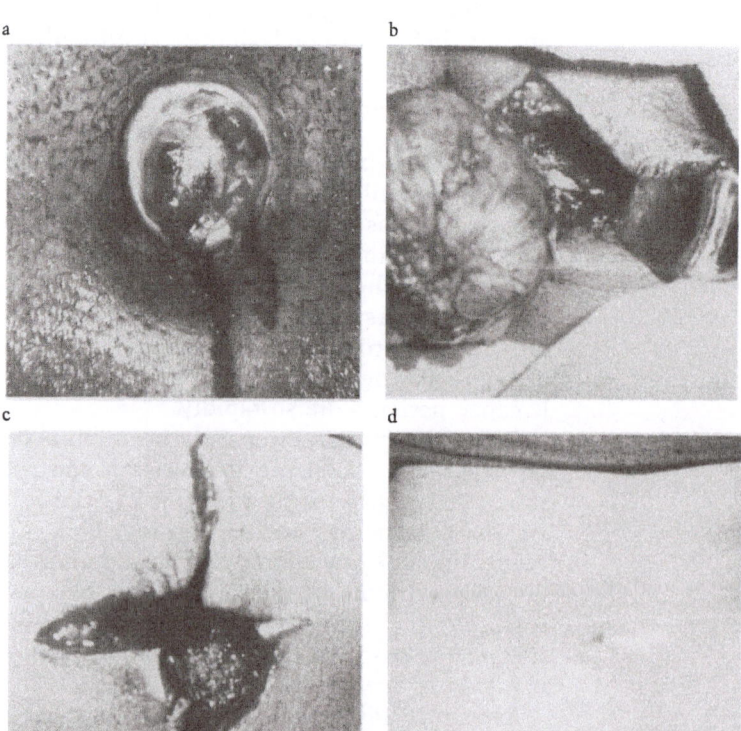

Figure 10.1
The ulcerated tumor mass (a) measured 4 cm. With electrocautery surgery (b), the tumor mass is completely excised under local anesthesia and the residual cavity (c) granulates over a 4 to 6 week period to heal completely (d).

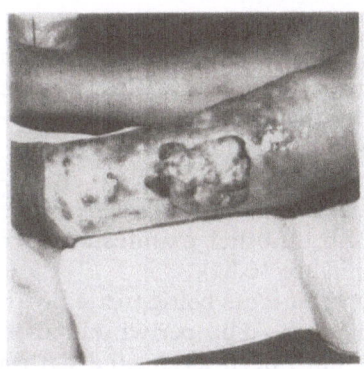

Figure 10.2
This cutaneous melanoma is superficial and exophytic with major secondary inflammation. Electrocautery surgery excised the lesion, leaving a residual granulating base, and precluded the necessity of amputation.

ample, herpes zoster infection or shingles, as it is known by the layman, is a viral infection of the nerves that generally manifests itself in a peripheral dermatome but occasionally along one of the facial nerves. This viral infection appears clinically as a vesicular rash with millet seed size blebs along the nerve distribution. The pain is typically radicular in type because it emanates from the spinal column down the length of the nerve. Occasionally, the pain will appear in the absence of a specific rash and is diagnostically confusing in this instance. However, in Hodgkin's disease and other lymphomas, the high frequency of secondary herpes zoster should always prompt shingles as a suspected diagnosis for any pain syndrome.

A composite of atypical pain syndromes including herpes zoster is detailed in Table 10.1. Referred pain has already been described; the most typical examples are the referral of pain from the distal extremities to the hip and from the diaphragm to the shoulder. The mechanism for referred pain is related to the multiplicity of afferent inputs which may traverse alternative neurologic pathways resulting in the transfer of the appreciation of pain from one site to another. Phantom limb pain represents a unique form of referred pain. Every patient who undergoes an amputation has phantom limb pain for some period of time. The development of phantom limb pain is caused by the pain that existed prior to amputation. The theory proposed is that a reverberating pathway develops in the central nervous system which has an automaticity that is cyclic. Relief of such pain is often accomplished by electrical stimulation of the peripheral nerve in the amputated limb. This therapeutic process interrupts the reverberation and breaks the cycle providing pain relief.

Pulmonary osteoarthropathy is an unusual manifestation of tumors, primary or metastatic to the lungs, in which a diffuse bone pain syndrome develops with tenderness along the long bones and in their periarticular areas. Elevation of the periostium is observed on radiographs generally but diagnosis may

Table 10.1
Atypical or Rare Pain Syndromes

Herpes Zoster (Shingles)	Viral infection of nerve root.
Referred Pain	Polysynaptic transmission to ancillary nerves.
Phantom Limb Pain	Central (CNS) mechanism involving reverberating circuits.
Drug Related Pain	Mechanism unknown but related to direct drug effect on the nerve sheath.

149

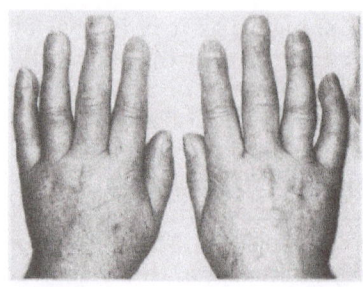

a

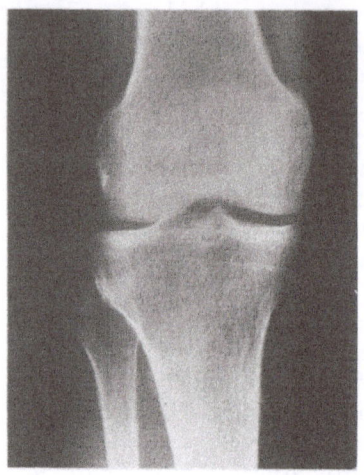

b

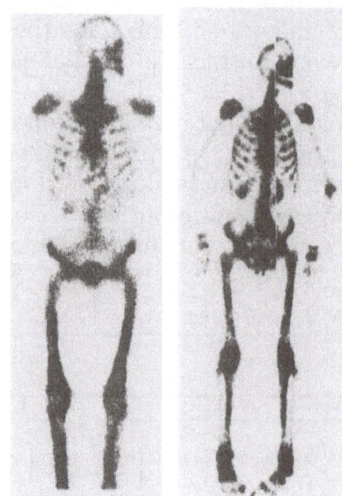

c

Figure 10.3
The syndrome of hypertrophic pulmonary osteoarthropathy is characterised by: (a) clubbing digits, and by (b) raised or elevated periostium in the long bone. (c) A radionuclide bone scan may demonstrate increased activity along the margin of the bones.

be made only by bone scan. The mechanism by which the pulmonary lung lesion results in secondary diffuse bone pain is unclear but may be related to a neuro-humoral substance released directly or stimulated by the tumor. Removal of the lung tumor, either by surgery or by other forms of treatment, results in acute and dramatic regression of the pain. An example of the particular bone scan and radiograph appearance is illustrated in Figures 10.3a & b. Another manifestation of the syndrome is clubbed digits (Figure 10.3c).

A pain syndrome may develop in patients being treated with specific types of drugs for malignancy. The periwinkle alkaloids and especially vinblastine have been associated with a neuro-myalgia syndrome which recedes in time following drug withdrawal but which may persist for perhaps weeks. Localizing jaw pain is similarly characteristic of the toxicity secondary to the vincristine-vinblastine drugs. Presumably, as new drugs are developed, particularly those related to plant alkaloids, the possibility of drug induced pain syndromes may become a more frequent problem.

Minor and rare forms of atypical pain syndromes include the localized pain secondary to alcohol ingestion observed in patients with Hodgkin's disease, and an unusual form of neuropathy related to nerve compression meralgia parastetica. The latter syndrome develops as a consequence of compression of the lateral femoral cutaneous nerve as it passes over the iliac crest, and may be observed in patients who are chronically confined to bed and lying on one side, with major weight loss and secondary loss of padding protection for the nerve.

6.0   Management of Pain Syndromes

The previous sections allude primarily to identification and characterization of the pain syndrome in order to more specifically delineate those pain mechanisms which are secondary to local factors. These local factors in turn may be relieved by surgery or radiation therapy. If they are related to non-tumor induced mechanisms, such as benign inflammatory conditions, they can be relieved by the appropriate therapy for those conditions. However, the following two common developments may necessitate consideration of analgesic drug therapy: (1) resistance to local mechanisms of management such as radiation therapy or surgery, and/or (2) systemic or diffuse nonlocalization of pain. Analgesia, meaning relief of pain, may be accomplished with a large variety of drugs. The drugs are listed in increasing strength, narcotic component, and potential for addiction in Table 10.2. In employing such drugs, two important rules may be followed. First, it is generally useful to employ the less potent drug initially to reduce the potential for addiction and secondary somnolence, and then to advance to the more potent drug as the pain becomes more resistant. Secondly, analgesia should be employed prophylactically in a regular schedule in order to diminish or prevent the redevelopment of pain. This latter guideline has evolved as a consequence of the readily established fact that intense pain requires

150

Table 10.2
Analgesic Drugs Organized by Increasing Effectiveness Levels When Employed as a Single Agent in Optimal Dose and Schedule.

Level	Drugs	Dose	Schedule	Comment
I	Aspirin, Phenace-tin, Acetomena-phen	2 tabs	q4–6h	Non-addicting
II	Percodan	1 or 2 tabs	q4–6h	Addicting
III	Demerol	100 mg	q4–6h	Ineffective orally at <100 mg
	Dilaudid	4 mg	q4–6h	
	Morphine	10 mg s.q.	q2–4h	
	Heroin	Variable	Variable	Unavailable
IV	Brompton Solution	See text	See text	See *Management of Pain Syndromes*

a higher dose level of narcotic than the same pain which may be anticipated and interdicted when it is at a lower intensity.

The mechanism of action of the analgesic is mediated through a suppression of the central appreciation of pain. As a consequence, ancillary effects of such drugs are somnolence, disorientation, confusion, and lethargy.

Psychotropic drugs are often employed for pain control (Table 10.3). In addition to physical pain, the patient may experience anxiety and mood alterations. The patient's psychologic

Table 10.3
Ancillary (Non Narcotic) Non Analgesic Drugs

	Use
Phenothiazines	May be synergistic when used concomitantly with narcotics to augment pain control
Antihistamines	Same
Tranquilizers	For patients in whom anxiety decreases the pain threshold
Sedatives or Soporifics	For pain induced insomnia
Mood Elevators	For patients in whom depression reinforces the pain syndrome

Table 10.4 Brompton Solution #1	
Morphine Sulfate	0.5 Gm.
Cocaine HCl	0.5 Gm.
Citric Acid	2.0 Gm.
Propylene Glycol	100.0 ml.
Alcohol U.S.P. 95%	300.0 ml.
Sorbitol Solution 70%	250.0 ml.
Saccharin Sodium	0.5 Gm.
Berry-Citrus Blend	4.0 ml.
Water q.s. ad.	1000.0 ml.

Dissolve Morphine Sulfate, Cocaine HCl and Citric Acid in 100.0 ml. of water.

Add with mixing Sorbitol Solution, Propylene Glycol and Alcohol in that order.

Dissolve Saccharin in 50.0 ml. of water. Add with constant stirring to above solution.

Adjust to 900.0 ml. Assay Morphine and Cocaine. Add flavor.

q.s. ad. with water to make 1000.0 ml. Filter.

status may intensify the pain, may provoke and promote withdrawal, and may potentiate the somnolence syndrome. Therefore, two therapeutic methods have evolved. First, depressed patients are treated with mood elevators and anxious patients with relaxants, such as tranquilizers. With the use of these drugs, the dose of the narcotic can often be reduced. Secondly, the narcotic may be used in conjunction with an antihistamine or a phenothiazine. This combination of narcotic plus supplement is synergistic in promoting pain control and will minimize drug dependence; however, it may cause significant drowsiness. The phenothiazine may also adversely affect the brain structures that control bodily movement, thereby resulting in extrapyramidal dyskinesia. Another complicating effect of the narcotics is the development of major constipation which may necessitate the regular use of cathartics. This subject is reviewed in the section on gastrointestinal effects.

One specific type of analgesia is provided by the Brompton Mixture, an oral narcotic preparation administered often with a phenothiazine. The Brompton Mixture (its formula is included in Table 10.4) consists of cocaine, alcohol, and morphine. This solution induces a controlled euphoric state, alleviates pain, and has the major advantage of not clouding the sensorium and inducing somnolence. Furthermore, the liquid oral preparation is easily administered and palatable. The use of this prophylactically on a regular schedule, again in contradistinction to the "demand" administration of narcotics, may return terminal cancer patients to a functional status for a meaningful period of time.

An important component of pain control is the appreciation of the placebo effect in pain management. More than 50% of patients may respond to a placebo which in Latin translates as "I will be pleasing." The placebo effect is defined as pain control induced by an inert substance. Patients whose pain is relieved by a sugar pill or water injection do in fact experience true pain control. Patient groups which demonstrated a higher response to placebo include the highly educated and professional personnel. Patients resistant to placebo were people who either had a low educational background or who were unskilled workers. The patient who benefits from a placebo is therefore likely to be an independent individual who enters into a dependent relationship as a consequence of a chronic illness. The mechanism of response to a placebo may be the autohypnosis phenomenon discussed in an earlier section.

7.0

Neurosurgical Approaches to Pain Management
Neurosurgical approaches to pain management are primarily employed in patients with lateralized pain due to cancer—that is, pain restricted to the right or left sides of the body. Lateralization is important in pain evaluation because if a neurologic motor deficit follows the procedure, the opposite side can function normally and compensate for the motor loss. Lateralization therefore dictates which side of the body is to be used in the surgical procedure.

Table 10.5
Neurosurgical Approaches to Pain Management

Peripheral Nerve Interruption
Sympathectomy or Ganglion Block

Dorsal Rhizotomy

Cordotomy
  Open Procedure (Laminectomy)
  Percutaneous (C_1–C_2 level)

Although a number of neurosurgical procedures may be employed in management of pain (Table 10.5), the most useful procedure is endolateral cordotomy to interrupt the spinothalamic tract which carries pain impulses to the brain. The cordotomy may be approached by an open surgical technique or alternatively by a percutaneous cordotomy when pain is derived from the upper limbs. The cordotomy is performed under local anesthesia by insertion of an electrode into the anterolateral column of the cervical spinal cord and administration of a radio frequency impulse through the electrode to destroy in a focal manner the local neuronal structure. Neurologic function and particularly pulmonary function are monitored closely. The recovery of motor function is common, but occasionally pulmonary compromise is compounded by local pulmonary disease and respiratory failure ensues.

Bilateral cordotomy is rarely considered for pain control because of the potential complication of loss of bowel and bladder sphincter function. In addition, high cordotomy may be associated with respiratory depression; bilateral procedures should not be used for patients with diminished pulmonary reserve.

An illustration of the various neurosurgical approaches is depicted in Figure 10.4. A sympathectomy is performed primarily for intra-abdominal pain. The neurosurgical approach to malignant pain is guided by the principle of operating as low as possible to avoid any unnecessary neurologic deficit in terms of motor or sensory function. The most peripheral procedure is a rhizotomy which is an interruption of the nerve root proximal to the dorsal root ganglion, thus affecting all afferents. Peripheral nerve interruption is analogous to a rhizotomy although the former is more localized and accessible to surgical approach. Cordotomy involves interruption of the spinal thalamic tract within the anterolateral column as indicated in Figure 10.4.

Neurosurgical procedures are employed only when the patient has exhausted specific local measures and analgesic medication. The major determinant in the consideration of a surgical pain operation is the balance of morbidity (of possible neurologic functional loss) against the life expectancy of the patient.

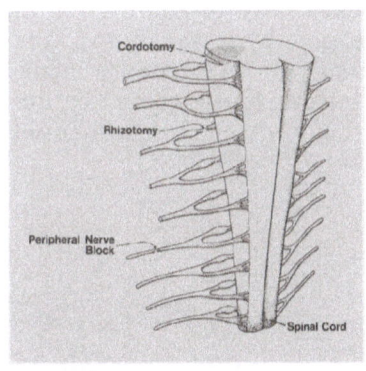

Figure 10.4
The neurosurgical techniques for controlling pain are: peripheral nerve sectioning, dorsal rhizotomy, and spinal cordotomy. They are generally employed in chronologic sequence.

8.0

Unusual Methods of Pain Management

The pathophysiology of pain suggests that the sensation of pain, relative to thresholds of detection as well as to intensity of pain appreciation, is located centrally in the cerebrum. This localization is the basis for the effectiveness of the majority of the analgesic drugs as indicated previously. Mind conditioning has been demonstrated to affect physiologic functions for some time and hypnosis has been employed successfully in the amelioration, for example, of central vomiting (see Chapter 8). Substantiating evidence for the potential for mind control of pain is provided by the numerous examples of placebo drugs. Mind conditioning represents a form of self-hypnosis. It is a learned process that can be employed in a variety of ways to assist the cancer patient. The actual application of hypnosis to pain control in the cancer patient has been explored casually but never in a controlled clinical fashion.

Acupuncture may be distinguished from hypnosis, at least by trained acupuncturists, and again has been informally demonstrated to be effective in pain control. All such methods, however, should only be employed in conjunction with standard forms of treatment and should never be the sole source of pain modulation.

9.0 **Summary**

The management of pain is of major importance to the cancer patient. The responsibility of the health care team is to categorize the pain and to direct treatment of the tumor when a specific tumor related etiology has been established. Characterization and localization of the pain will define the therapeutic approach. Thus, diffuse pain will be managed by systemic therapy and local pain will be managed by surgery or radiation. The use of analgesics in appropriate combinations should be employed prophylactically rather than therapeutically. A pain-free state should be achieved whenever possible, especially for patients with chronic malignant pain. Neurosurgical approaches should be used only in those circumstances where pain is not controllable by lesser means. The dictum to be emphasized, particularly in patients with incurable and terminal disease, is that narcotic use should be liberalized without regard for addiction or other inconsequential adverse effects.

11 Management of Constitutional Symptoms

1.0

Constitutional Symptoms

Constitutional symptoms are general as opposed to local symptoms which affect the cancer patient's entire system. This "systemic symptomatology" is most often a complex of multiple symptoms which are interrelated, which develop sequentially, and which reinforce one another. For example, physical weakness due to illness may cause a depressed self-image and promote withdrawal which in turn brings on somnolence and disinterest in normal activities, such as ingesting food. A vicious cycle develops which if untreated can have a serious impact on a patient's physical and emotional well being.

Unfortunately, these symptoms and the constitutional syndrome representing them are often disregarded by the health care team. They are accepted as inevitable aspects of cancer which need only marginal treatment. The diagnostic and therapeutic approach to patients suffering from the constitutional syndrome, however, should be aggressive and optimistic. The constitutional syndrome is a deterioration of the essential or primary drives in people. Food, rest, sex, and productivity are motivational forces in our lives; in illness, these crucial components are the first to be affected. The potentially reversible causes of the syndrome can be treated specifically but even in the absence of a specific diagnosis, symptomatic therapy may be helpful. Table 11.1 describes the individual components of the syndrome relative to the normal activity and the therapeutic options in general terms.

We will now break constitutional symptoms down into specific categories and discuss appropriate ways of treating them.

Weakness may be described in various ways by the patient. Lassitude, fatigue, and shortness of breath are all descriptive terms. The degree of weakness may be evaluated by performance status or by completion of daily activities. Inability to climb stairs or to do daily housework are meaningful indications of weakness. A useful means of quantitation is a grada-

Table 11.1
Symptom Complexes of the Constitutional Syndrome of the Cancer Patient

Abnormal Symptom	Normal Activity	Most Common Causes	Treatment
1. Weakness	Motivation, Productivity	Metabolic, Depression Therapy (Radiation)	Antidepressants; Correct Metabolism
2. Lethargy & Somnolence	Alertness	Medication	Reduce Dose of Medication
3. Insomnia	Rest	Cortico-Steroids	Sedatives, Hypnotics
4. Impotence or Decreased Sexual Interest	Coitus	Metabolic	Ineffective
5. Anorexia	Food Ingestion	Idiopathic, Metabolic	Hyperalimentation

tion system of performance status relative to normal daily activities. It is set up as follows: Grade 0, working full time; Grade 1, working part time; Grade 2, unable to work but completely ambulatory; Grade 3, bedridden less than 50% of the time; Grade 4, bedridden more than 50% of the time.

An appropriate diagnostic workup is important to eliminate specific metabolic causes of the weakness syndrome. Common causes of weakness are: hypercalcemia related to osseous metastases or to tumor secretion of parathormone, hypokalemia due to excessive diuretic use, protracted vomiting or corticosteroid therapy. Another common cause is the generalized weakness that occurs because of radiation therapy known as radiation sickness. Here, the patient experiences anorexia and intermittent nausea. These symptoms may develop with even small doses of radiation therapy to localized areas outside the abdomen.

The most common cause of weakness, however, is the depression withdrawal syndrome which becomes a cyclic reinforcement phenomenon with depression leading to withdrawal and sensory deprivation resulting in weakness and muscle atrophy. The secondary physical manifestation leads to a reinforcement of depression and further withdrawal. This cycle must be interrupted by an optimistic, positive attitude by the health care team and family. Often counseling is salutory, and psychotrophic drugs, such as mood elevators or antidepressants, may be effective.

Somnolence. The somnolence syndrome represents a pathologic need for sleep and is most often iatrogenically caused

158

by excessive use of narcotics or sleeping medication and soporifics. It is particularly apparent in patients who have an altered metabolism due to cachexia and hepatic or renal dysfunction. The sensitivity to such drugs is often increased because of the decreased metabolism of the drug and prolonged blood levels. Therapeutically, it is important to correct the iatrogenically induced somnolence syndrome by reduction of a dose of soporifics or tranquilizers. Somnolence may represent an extreme form of depression and withdrawal. A continuum or a spectrum of physical lassitude may be observed from weakness to somnolence. If the specific etiology of somnolence is not organic, it is important for the health care team to establish performance goals for the patients and to encourage any form of activity on a daily basis. The use of pharmacologic stimulants such as amphetamines for these patients is not suggested because the altered metabolism of such drugs may result in major adverse effects and the therapeutic usefulness is marginal.

Somnolence, as suggested previously, is one end of the spectrum from weakness through lethargy to the soporific state and finally coma. Thus, the metabolic and organic causes of somnolence are precisely similar to those related to weakness. In addition, however, the somnolent patient develops neurologic components of dysfunction which are nonspecific but which reflect a distorted state of mentation. Disorientation, confusion, and erratic behavior may be evident and are rarely focal neurologic signs. The somnolent patient should be evaluated for the possibility of an intracerebral process as well as the metabolic causes of encephalopathy. In addition, the endocrine and paraneoplastic disorders associated with malignancy must be considered within the metabolic etiologies. Most commonly, however, iatrogenic causes brought on by the excessive use of psychotropic drugs, particularly narcotics, tranquilizers, and soporifics, are the primary etiologies of somnolence.

Impotence. Impotence in the male or decreased sexual interest in the male or the female is a common symptom for patients with metastatic cancer. It is appropriately included in the constitutional syndrome because it genuinely interferes with a primary human need and a life style component. With few exceptions, the specific etiology of impotence or decreased sexual interest cannot be established. Specific organic causes of impotence are diabetes mellitus, vascular insufficiency, and primary gonadal dysfunction. The last etiology is extremely uncommon with the exception that patients with prostatic cancer treated by orchidectomy or estrogen therapy are uniformly impotent and are also disinterested in sexual activity. Finally, pelvic surgery (for patients with bladder, rectal, prostate, or testicular cancer) may compromise the nerve supply to the genitalia, resulting in impotence. Most commonly, however, impotence is simply a reflection of generalized illness, secondary withdrawal, and depression. This symptom is distressing and should be discussed with patients frankly.

159

Anorexia. Anorexia may be described as decreased appetite or actual food intolerance with strong negative reactions to specific or general food intake. This symptom is a crucial constitutional symptom for the cancer patient and results from specific metabolic alterations brought about by the tumor or a secretory product of the tumor which directly affects appetite. A decreased appetite is commonly associated with metastases to the liver but this reaction is not universal. Patients with ascites or large abdominal mass lesions may develop early satiety and decreased food intake because of compression of the stomach, but hunger persists. Many ancillary effects of cancer lead to anorexia including decreased taste and smell, decreased salivation, and fear of abdominal pain or defecation. However, the palatability of food is the major contributing factor to anorexia.

Anorexia and decreased food intake adversely affect the response of the tumor to therapy as well as the tolerance of the patient to therapy. Furthermore, the engagement in an activity which may have been a major source of enjoyment suddenly becomes a negative experience. The evaluation of and approach to anorexia is described in the chapter on nutrition.

Insomnia. The antithesis of lethargy and somnolence is insomnia and the patient who is unable to sleep suffers as much as the patient who sleeps excessively. Insomnia may be defined as constant sleep interruption because of organic symptoms, such as bone pain, micturition (nocturia), or shortness of breath (orthopnea). Non-disease-related insomnia is more common. Therapy in these instances should take into account the possibility of a patient's morbid fear of nightmares or of death while asleep. Non-disease causes of insomnia must be overcome by reassurance as well as by appropriate tranquilizers and sedative medication.

An alteration in the normal circadian sleep pattern is common in patients with malignant disease and the cause or specific pathophysiology is unexplained. Corticosteroids specifically may alter the circadian rhythm if employed in a dose and schedule which results in high steroid blood levels in the late evenings. Corticosteroids for such patients should be administered no later than mid-afternoon.

Soporifics can be effective for the insomniac. The judicious use of sedatives is often crucial to patient comfort because the patient who is fatigued from lack of sleep or from excessive sedative medication is in effect debilitated. In general, the multiplicity of types of sedatives offers a wide range of potentially effective drugs and patients may be totally unresponsive to one while being very sensitive to another. Thus, it is useful to employ the sedative preparations in sequence in order to identify the most useful drug for an individual patient. This approach is preferable to increasing the dose of one specific preparation. The barbiturates, in particular, may cause profound respiratory depression, possible pulmonary aspiration, or more commonly, profound post-barbiturate hangover with increased dose.

Differential Diagnosis

Although constitutional symptoms are nonspecific, each may be indirectly or directly related to the tumor or to the treatment for the tumor. When the constitutional symptoms complex is recognized, each symptom should be evaluated individually to determine if it is etiologically related to the tumor, the tumor therapy, or to an incidental disease. The specific treatment of the symptom is dependent upon establishing a specific etiology. Alternatively, symptomatic therapy of a nonspecific type may be employed with much less expectation of success. For example, the patient with "nonspecific" fatigue syndrome is unlikely to benefit from motivational therapy, but if the fatigue is related to decreased thyroid function, providing thyroid supplements may reverse the syndrome completely.

The differential diagnosis of the components of the constitutional symptoms is obviously extensive but if one focuses on the potentially correctible causes associated with cancer, a well defined and relatively narrow list may be established (Table 11.2). The metabolic causes of weakness most treatable are decreased potassium or increased calcium levels. Hypokalemia may be associated with diuretic therapy, with excessive vomiting or possibly with an excessive cortisone syndrome. Hypercalcemia may be secondary to tumor secretion of PTH or due to bone metastases. In both electrolyte syndromes, weakness may be a prominent symptom, but constipation, polyuria, and anorexia may also be observed. Hyponatremia (decreased sodium) may cause a variety of ancillary symptoms including muscle cramps as well as lassitude and fatigue. Hyponatremia may be caused by excessive vomiting or diuretic use. An unusual cause of hyponatremia is the syndrome of inappropriate ADH secretion which results in water intoxication. Hepatic and renal insufficiency are common metabolic abnormalities which may be associated with correctible surgical lesions, such as obstruction of the biliary tree or the ureters. Pulmonary insufficiency may develop as a consequence of the replacement of the pulmonary parenchymal with a tumor but more commonly is secondary to ancillary chronic

---

Table 11.2
Differential Diagnosis of the Symptom Complex within the Constitutional Syndrome

1. Metabolic Causes	Hyponatremia
	Hypokalemia
	Hypocalcemia
	Hepatic Insufficiency
	Pulmonary Insufficiency
	Renal Failure
2. Iatrogenic Causes	Neuropathy 2° periwinkle alkaloids
	Myopathy 2° cortico-steroids
	Radiation Therapy

pulmonary disease, such as emphysema. Treatment related or iatrogenic causes of weakness are generally due to drug therapy which directly affects nerves or muscles. The periwinkle alkaloids (vincristine) have a direct effect on the nerves and cause neuropathy. More commonly, the use of corticosteroids leads to a proximal myopathy which manifests itself as an inability to climb stairs. Radiation therapy often results in a nonspecific fatigue syndrome characterized by somnolence, anorexia, intermittent nausea, and weakness.

All of the constitutional effects induced by therapy recede with drug withdrawal or discontinuation of radiation, but the effect may last from weeks to months. The symptoms secondary to metabolic abnormalities, however, recede promptly with reversal of the organ dysfunction or correction of the electrolyte imbalance. The importance of establishing a specific etiology for constitutional symptoms is exemplified by the following case.

A 65 year old man had a 4 week history of progressive weakness. The weakness had progressed to the point of inability to rise from bed and he was somnolent and disoriented. In the past, the patient was known to have had cancer of the prostate gland and diffuse bone metastases. Examination revealed a large distended bladder and by rectal exam a firm, hard prostate was palpated. Laboratory examination revealed a blood urea of 150. At this juncture a decision to accept the diagnosis of metastatic cancer was considered. However, since a possible source of all the symptoms was the localized obstruction with secondary renal failure, it was decided to attempt catheter drainage.

The patient had a catheter inserted and subsequently a transurethral resection identified carcinoma in the prostate. Eventually the BUN returned to normal and the patient returned to a normal life style and was completely active. The point of this example is that by identifying the cause of simplistic generalized symptoms, the health care team may return what appears to be a totally incapacitated patient with limited life expectancy to a normal existence.

3.0

General Therapeutic Options
In the absence of specific organic causes or of an identifiable metabolic etiology for the five individual components of constitutional symptoms, the therapeutic options are limited. The four broad categories of etiologic mechanisms for the constitutional symptoms are: (1) metabolic causes; (2) iatrogenic causes; (3) idiopathic causes; and (4) psychogenic causes. The first two may be regarded as the reversible or potentially treatable causes and are listed specifically in Table 11.2. The idiopathic category, unfortunately representing the majority of patients, reflects our present inability to identify a specific cause for many of the symptoms, such as anorexia. The psychogenic category is basically an etiologic diagnosis of exclusion but offers some therapeutic options. Depression, anxiety, and

withdrawal may be managed with psycho-active drugs and psychotherapy. With these means, a reversal or amelioration of the constitutional syndrome is achievable.

Hormone therapy is commonly employed in patients with malignancy, particularly in the terminal stages. Corticosteroids are especially popular because they create euphoria and stimulate the appetite. Such salutory effects, however, are generally temporary and the subsequent development of side effects from steroids often supersedes their potential benefits. Administration of sex hormones has also been employed but scientific evidence of a patient's benefit from them is difficult to establish. Androgens are well known to have an anabolic effect on protein synthesis, to stimulate the erythropoetic system, and to reverse anemia sometimes. Obviously, the use of androgens is not acceptable for women and is often undesirable for men. The androgens may stimulate an occult prostatic cancer or may create sexual impulses which are disquieting for elderly men.

Psychological supports are the essential ingredient in therapy for patients experiencing constitutional symptoms when an organic cause, specific metabolic, or iatrogenic etiology cannot be established. Effective psychotherapy is dependent upon an established relationship between the health care team and the patient. The health care team must constantly encourage patients to develop new experiences and to be sensitive to their family and environment. Sustaining a patient's involvement in life and preventing the physical and emotional withdrawal syndrome may be accomplished by the simplistic tactic of daily activities (occupational therapy) and by the health care team's setting of realistic goals of accomplishment for the patient.

Newer forms of therapy for constitutional symptoms are being explored and are in various stages of development. The use of lysergic acid (LSD) derivatives which create a controllable euphoria and maintain an acute sensitivity with consistently positive overtones to environmental stimuli is being explored in some centers. Similarly, the use of tetrahydrocannabinol or marijuana in a controlled fashion may promote the positive interaction of patients with their families. It may also improve their self-image and thereby prevent withdrawal and a morbid focusing of attention on constitutional symptoms.

Section IV   Special Support Therapies

12   Management of Nutrition

1.0

Introduction

Cancer can be a wasting disease of multiple causation. If it progresses unchecked, malnutrition is an inevitable consequence. Malnourishment may arise because loss of appetite reduces calorie and protein intake with resultant weight loss. The anorexia of cancer may be due to either specific metabolic effects of the tumor or altered cell metabolism wherein there is nutrient waste. Antineoplastic therapy with its side effects of anorexia, malabsorption, nausea, vomiting, and diarrhea is often a potent cause of cancer cachexia.

We must realize that malnutrition is an important factor in the morbidity of cancer. It is important to recognize such actual or incipient debilitation and to define and characterize its particular features so that a metabolic and nutritional profile of each cancer patient can be achieved. With such information the most effective therapeutic plan, a "nutrition prescription," can be written by using recently refined techniques of enteral and parenteral hyperalimentation, which will be discussed later. Dramatic improvement ("palliation") can occur with prevention of cancer cachexia. Nutritional support can enormously aid the ongoing specific antineoplastic therapy of the surgeon, radiotherapist, or medical oncologist. However, it should not be considered in isolation but must become part of the broad treatment plan in close synchronization with other specialties. Of course, the patient's condition must be regarded as an integral part of the therapy (Figure 12.1).

For example, "force feeding" by hyperalimentation allows one to reverse the downhill course of the nutritionally depleted cancer host (patient). Such therapy encourages anabolism, the synthesis of essential body protein, so that patients have improved wound healing, tissue repair, and immune function. Patients with adequate diets become stronger, and their morale improves. Exercise, which further promotes anabolism, must be encouraged. Positively motivated patients are better able to withstand the stress of cancer treatment: there is a superior

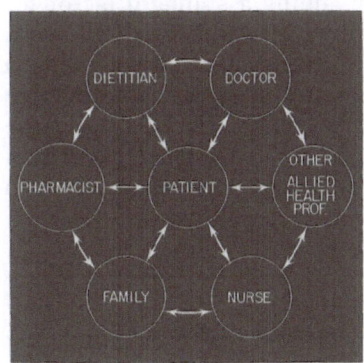

Figure 12.1
Hospital-Patient Supportive Interrelationship

response to surgical intervention as well as an increased tolerance to radiotherapy and chemotherapy.

Nutritional support of the cancer patient is certainly possible and worthwhile, and has a vital sustaining role as an adjunct to other specific cancer treatment. Furthermore, the use of various specific nutrients has been shown to be beneficial as "placebo" medicine—a not to be ignored aspect of treatment which must be used increasingly until public health measures in preventive medicine produce a reduced incidence of carcinoma, and/or improved antineoplastic remedies are developed.

It should be stressed that sophisticated nutritional support of the cancer patient who is terminal, and of those in whom other therapeutic measures have been withdrawn, should not be countenanced. However, we can help to improve tolerance of palliative treatment, aiming for a longer hospital-free course and improved quality of remaining life.

2.0

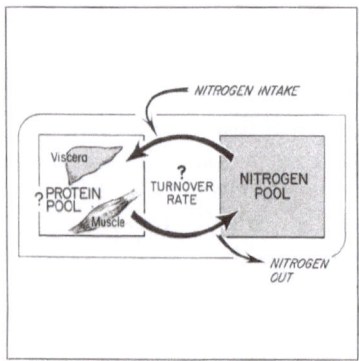

Figure 12.2
Limitations of Nitrogen Balance

2.1

Protein and Energy Metabolism
The body requires these basic nutrients:

(1) Carbohydrate, mainly as an energy source
(2) Fat, mainly as an energy source
(3) Protein for building and bodily repair
(4) Vitamins, to support enzymatic function and oxidation
(5) Minerals
(6) Water and electrolytes for homeostasis

In normal human metabolism, intake and output are equal: *catabolism* (tissue breakdown) equals *anabolism* (tissue synthesis) which yields a steady weight and a roughly zero nitrogen and energy balance (Figure 12.2). Determination of the nitrogen balance is the only way to evaluate the effect of diet from day to day.

Calculating Nitrogen Balance
The nitrogen balance is the subtraction of nitrogen output from nitrogen intake. Intake can be calculated from the dietary input of protein divided by 6.25, as protein is 16% nitrogen. Most of the daily nitrogen output is in the form of urea nitrogen excreted by the kidneys. By adding a standard factor for non-urea nitrogen in the urine plus losses through the skin and feces (Table 12.1), an adequate clinical estimate of nitrogen balance can be made. Cancer patients in significant negative nitrogen balance of 4 to 8 grams per day will rapidly become "protein malnourished." The goal for a depleted patient in terms of nitrogen balance should be a positive balance of 4 to 5 grams each day.

Table 12.1
Calculation of Nitrogen Balance

$$\text{Nitrogen Balance} = \frac{\text{Protein Intake}}{6.25} - (\text{urinary urea nitrogen} + 4)$$

166

Energy balance is assumed if weight remains steady; positive energy balance is witnessed by storage of extra energy as fat. If, on the other hand, energy intake is insufficient to meet physiologic demands, the body sacrifices its own energy stores. With enough glycogen (carbohydrate) in the liver for only 12 to 24 hours on a carbohydrate free diet, body fat and protein begin to be sacrificed. An important point here is that rapid weight loss is associated proportionately with more protein breakdown, as demonstrated by increased urine urea nitrogen. Up to 50% of rapid weight loss is due to diminished body protein. (In contrast, in a non-stressed, starvation-adapted state, humans can subsist mainly on their fat stores with much reduced protein losses.) The cancer patient has lost this ability to adapt to starvation and has rapid weight loss as well as a negative nitrogen balance. Thus, the patient's loss of appetite combined with his altered metabolism leads to a sacrifice of the functioning protein to the support of preferential tumor growth.

Usual protein requirements can be met by the intake of about 100 grams of protein in addition to 2,000 non-protein calories in order to spare use of protein as energy fuel. The energy content of the various available sources is:

(1) Fat	9 kilocalories per gram
(2) Alcohol	7 kilocalories per gram
(3) Carbohydrate	4 kilocalories per gram (oral) 3.6 kilocalories per gram (intravenous)
(4) Protein	4 kilocalories per gram

It can be seen that fat is the most concentrated energy source available to humans, although carbohydrate is the major supplier of energy in a typical American diet. Given the poor appetite of many cancer patients, high-density feeding above 1.5 calories per cc (cal/cc) is most desirable. Micro-nutrients of vitamins and minerals are best provided in an inexpensive supplement. Prevention of weight loss often depends on the number of bites or swallows the patient can tolerate. Meat, potatoes, and desserts are more important than such classic food groups as fruits or vegetables. This preference is in contrast to what would be key goals in the "prudent diet" for preventive health.

3.0 **Cancer Cachexia**

Cachexia is a specific syndrome, characteristic of patients with malignancy, particularly in more advanced and metastatic forms of the disease. Its features are:

(1) Weakness
(2) Loss of appetite
(3) Depleted and/or altered host body compartments, e.g., weight loss
(4) Hormonal aberrations
(5) Progressive loss of vital functions (leading to death)

167

Table 12.2
Causes of Anorexia

1 Alterations in taste and smell

2 Products of metabolism

3 Hypothetical tumor toxins

4 Appetite center depression

5 Psychological factors

6 Inability to eat—hypophagia

Cancer patients' weakness reflects depleted nutritional status and loss of muscle mass, compounded by disuse atrophy. There is a decline in patients' morale as they find themselves unable to perform even simple tasks without marked effort.

The anorexia of cancer has many causes (Table 12.2), but they all result in reduced, insufficient caloric intake in a patient who has to support not only his or her own energy and protein needs, but also those of an undisciplined tumor. This tumor may compete preferentially for available substrate—that is, energy and protein.

An altered sense of taste and smell often victimizes cancer patients. It is well recognized that these patients may develop a marked aversion to meat: studies have correlated this phenomenon with a reduced oral threshold for urea, which has an unpleasant taste. Accordingly, patients will refuse high urea content food, such as red meat. Patients with head and neck cancer who are receiving radiotherapy may develop stomatitis with much reduced taste and smell sensibility amounting to "mouth blindness," a potent cause of anorexia. Lack of appetite may also result from psychological factors, such as depression and apathy. That cancer itself has a direct anorectic effect, perhaps at a hypothalamic level, is as yet undemonstrable; but with successful treatment of the tumor, anorexia can be reversed.

Other causes of a reduced food intake can be obstructions caused by upper gastrointestinal neoplasms, such as esophageal carcinoma. Malabsorption, a prominent feature of pancreatic neoplasms and small bowel lymphomas, is also a contributing factor.

Hormonal abnormalities can occur in cancer arising in nonendocrine tissue: lung cancer is a noteworthy example, where ADH- or ACTH-secreting tumors may occur. A different example is choriocarcinomata which produce human chorionic gonadotropins (HCG). The HCG has a TSH-like effect. It increases thyroid activity and the metabolic rate, further contributing to cachexia. Moreover, weight loss can be camouflaged by fluid retention, either in the form of massive ascites due to malignant peritoneal seeding from an ovarian or gastric carcinoma, or in the form of peripheral edema related to the hypoalbuminemia of malnutrition and anemia.

It is also probable that cancer exerts some specific metabolic action on the host, for the cachexia of malignancy occasionally seems out of all proportion to the bulk and anorectic stimuli of some tumors. For instance, in oat cell lung carcinoma, a small confined lesion may be associated with gross muscle wasting and weight loss.

Possible ways in which neoplasms may exert a prominent role in inducing protein-calorie malnutrition (marasmus) are:

(1) *Reduced host protein synthesis, tumor-induced*

(2) *Altered carbohydrate tolerance leading to a diabetic-like state*

(3) *Altered host protein homeostasis leading to negative nitrogen balance*

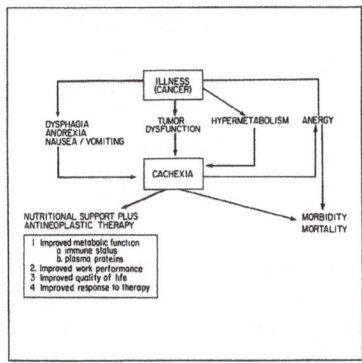

Figure 12.3
Pathogenesis of Cancer Cachexia

(4) Increased anaerobic glycolysis in the tumor
(5) Specific anorectic effect
(6) Preferential absorption of nutrients by the tumor
(7) Less efficient utilization of energy and nitrogen
(8) Reduced ability to adapt to the starved state; therefore inability to conserve vital body cell mass
(9) Increased metabolic rate (not clinically significant)

From this list, it can be seen that cancer may exert profound alterations in the metabolic homeostasis of the host—the syndrome of cancer cachexia. The patient becomes depleted nutritionally and has reduced immunocompetence, reduced resistance to infection, and an inability to respond well to stress. Without nutritional support, such a patient poorly tolerates other therapeutic endeavors (Figure 12.3).

4.0   **Nutritional and Metabolic Assessment**
Nutritional assessment is the first step in the treatment of the undernourished, i.e., one must delineate the nature and degree of nutrition and determine the energy (calorie) and nitrogen (protein) requirements of each individual. Functional status of the gastrointestinal tract should also be assessed. Repeat measurements should be made while treatment is continued to ascertain that the defined nutritional parameters are improving. These parameters are: (1) nutritional history, (2) objective clinical data, and (3) laboratory tests.

4.1   **Nutritional History**
It is important to define the severity and rapidity of weight loss, the patient's usual weight, and the presence or absence of anorexia, nausea, or vomiting. Questions must also be asked about eating habits, likes and dislikes, bowel habits, and tolerance of solids and liquids.

4.2   **Objective Clinical Data**
The patient's height and actual and usual weight are recorded. Then anthropometric measurements are made on the nondominant arm. Triceps skinfold thickness (TSF) is measured with Lange calipers as well as arm circumference at the same point. These are indices to arm muscle circumference (AMC), which is a simple way of estimating body muscle mass. The triceps skinfold gives an index of the patient's fat stores. Both measurements are compared with normal values and expressed as a percentage of the standard. A calorie and nitrogen count should be kept over several days to estimate dietary intake.

4.3   **Laboratory Tests**
The following investigative blood studies are performed: hemoglobin and hematocrit, total W.B.C., differential W.B.C., total lymphocyte count, reticulocyte count (%), BUN, glucose, electrolytes (Na, K, Cl, CO_2), SGOT, LDH, alkaline phosphate, bilirubin, albumin, transferrin, and total iron-binding capacity (TIBC). In addition, 24 hour urine samples are analyzed for: (1) urinary urea nitrogen: calculation of nitrogen balance and (2)

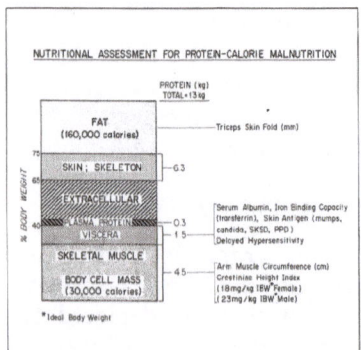

Figure 12.4

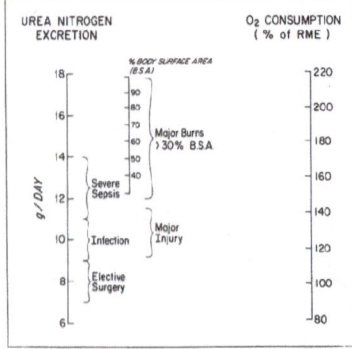

Figure 12.5
Calculation of degree of catabolism

creatinine: measure of renal function and calculation of creatinine-height index (C.H.I.). Finally, skin tests using recall skin antigens of mumps, candida, and streptokinase-streptodornase (SKSD) are performed.

The rationale for these numerous investigations and measurements can best be illustrated by reference to Figure 12.4. It can be seen that we assess:

A. *Fat stores* by triceps skinfold.

B. *Body cell mass*, the vital energy-consuming part of total body mass, by arm muscle circumference and creatinine height index. The latter is derived from actual urine creatinine divided by ideal urine creatinine times 100.

C. *Visceral protein status*, that part of the body cell mass exclusive of skeletal muscle and thus reflective of liver, gut, and smooth-muscle function, by serum albumin, transferrin, liver function tests (LFT's), hemoglobin (Hb), and reticulocyte count.

D. The *degree of catabolism* with some accuracy by the 24 hour urine urea nitrogen (Figure 12.5).

From Table 12.3, then, one can determine whether the patient has marasmus (protein-calorie malnutrition) as distinguished from kwashiorkor (protein malnutrition). Kwashiorkor was the name given by the Ga tribe near Accra, Ghana to "the sickness the older child gets when the next baby is born." This type is less easy to identify clinically as there may be little change in skeletal parameters, such as weight and anthropometrics, although visceral and immune parameters will

Table 12.3
Classification of Malnutrition

	% Ideal Weight	Creatinine Height Index (%)	Skin Test (mm)
A. Marasmus			
Moderate	60–80	60–80	
Severe	<60	<60	<5*

	Serum Albumin	Serum Transferrin	Total Lymphocyte Count	Skin Test (mm)
B. Kwashiorkor-like				
Moderate	2.1–3.0	100–150	800–1200	<5*
Severe	<2.1	<100	<800	<5*

Treatment of malnutrition requires characterization into two general categories that reflect their

1. Different consequences in terms of morbidity and mortality.
2. Different treatment modalities.
3. Different pathogenesis.

Furthermore, these *diseases* should be coded so that they will appear on hospital discharge summaries and in health planning statistics.

ICDA-International Classification[14]
Protein malnutrition–Kwashiorkor-like   No. 267
Nutritional marasmus   No. 268
Other nutritional disorders–unspecified   No. 269.9

*On all three skin tests.

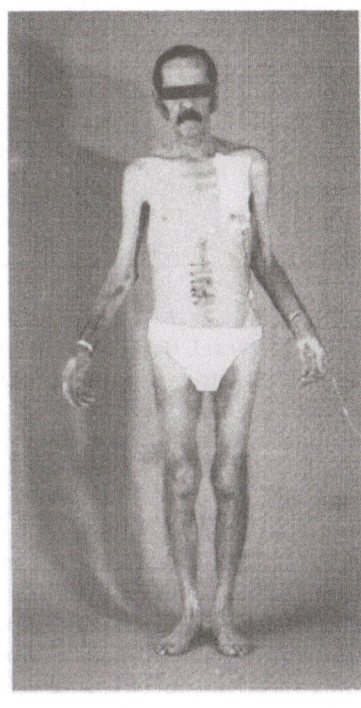

Table 12.4
Nutritional Therapy

A. Energy Requirements	Calories required (per 24 hrs.)
Type of therapy	
Parenteral anabolic	1.75 × BEE
Oral anabolic	1.50 × BEE
Oral maintenance	1.20 × BEE

B. Prescriptions for Anabolism*	Protein (g/day)	Calories (kcal/day)
Type of therapy		
Oral protein-sparing	1.5 × weight†	
Total parenteral nutrition	(1.2 to 1.5) × weight	40 × weight
Oral hyperalimentation	(1.2 to 1.5) × weight	35 × weight

*Levels of protein intake are to be adjusted according to blood urea nitrogen values and nitrogen balance.
†Weight = actual weight in kg.

be depressed. Figures 12.6 and 12.7 clearly illustrate the differences between marasmus and kwashiorkor. Cancer cachexia usually produces a marasmic patient. One should also be aware (especially in the surgical patient) of selective visceral protein depletion (low albumin, total lymphocyte count, and anergy), which has adverse influences on wound healing, tissue repair, and resistance to infection.

4.4

SUMMARY			
STANDARD PARAMETERS	90%	60-90%	60%
WEIGHT/HEIGHT		■	
TRICEPS SKINFOLD			■
MID UPPER ARM CIRCUMFERENCE		■	
MID UPPER ARM MUSCLE CIRCUMFERENCE		■	
ALBUMIN	■		
CREATININE HEIGHT INDEX			■
LYMPHOCYTE COUNT	■		
IRON BINDING CAPACITY/ TRANSFERRIN	■		
CELLULAR IMMUNITY SK/SD			■
OTHER CANDIDA			■
D.N.C.B.			■

Figure 12.6
Assessment of Marasmus Malnutrition

5.0

Calculation of Nutritional and Energy Requirements
Energy requirements are calculated by the Harris-Benedict Formula which estimates basal energy expenditure (BEE) by age, sex, and the usual weight and height indices. Anabolic requirements are expressed as a function of BEE (see Table 12.4). A rough calculation, however, can be made using the patient's weight: the energy requirement for anabolism is 40 to 45 kcals/kg/day; maintenance energy needs are 30 to 35 kcals/kg/day. Maintenance protein requirements are about 0.8 grams per kilogram per day (g/kg/day). However, it may be to the patient's benefit to increase his or her intake to 2 to 3 g/kg/day (or 20 to 25% of energy expenditure). Efficacy can be determined over the short term by calculating nitrogen balance plus estimated efficiency of dietary protein consumption: "net protein utilization" (NPU) (Table 12.5).

Techniques of Nutritional Support
Especially designed therapeutic feeding can be given enterally or parenterally. "Combined" treatment and "modular" feeding—which can greatly improve responses to specific nutritional prescriptions—will also be considered.

Enteral hyperalimentation refers to the use of the gastrointestinal tract to supply sufficient energy and nitrogen nutrients to reverse catabolism and produce anabolism. If the gastrointestinal tract functions, it should be used for feeding—even if for only a part of the overall effort. A variety of techniques are available: (1) oral feeding, (2) nasogastric tube feeding, (3) indwelling gastrostomy, and (4) indwelling jejunostomy. It may be apparent at once that the patient will be unable to tolerate

171

Table 12.5
Net Protein Utilization (N.P.U.)

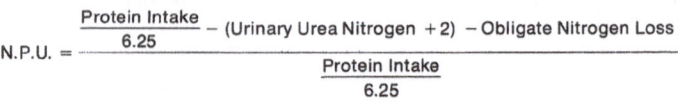

$$\text{N.P.U.} = \cfrac{\dfrac{\text{Protein Intake}}{6.25} - (\text{Urinary Urea Nitrogen} + 2) - \text{Obligate Nitrogen Loss}}{\dfrac{\text{Protein Intake}}{6.25}}$$

the necessary intake by mouth, and several days on caloric counts may reveal that the patient has overestimated his appetite. In practice, we have been greatly impressed by how many patients and doctors overestimate caloric intake and underestimate anorexia. The patient is often really subsisting on only fluids and low-protein foods.

The use of a nasogastric feeding tube should be considered when: (1) oral intake is insufficient, (2) the gastrointestinal tract is functional, (3) there are no high gastrointestinal tract fistulae, (4) the patient is cooperative and agreeable to trying tube feeding, and (5) the patient is not comatose and has a good cough reflex.

Silicone feeding tubes, which are mercury-weighted to aid passage down the esophagus and which can be visualized on x-rays of the abdomen, are well tolerated by the cancer patient.[1] Some patients are able to pass the tube down themselves each morning, and there is no interference with oral ingestion of food.

The enteral is preferable to the parenteral route as it is more physiologic and much cheaper and easier, and has far fewer potential complications. But some patients may not be able to tolerate nasogastric tubes, may have gastric stasis, or may have had problems with aspiration pneumonia. These patients are not good candidates for tube feeding. Before planning parenteral nutrition for such patients, however, one should take into account other techniques that are surgically available. Gastrostomy or jejunostomy tube feeding should be considered in nutritional support, particularly when long-term feeding is a possibility and a nasogastric tube is contraindicated. The patient, for example, may have an obstructed gastric outlet or may have undergone gastroenterostomy; here, the gut which is functional distal to the obstruction can be used with a jejunostomy tube with little risk of aspiration.

SUMMARY			
STANDARD PARAMETERS	90%	60-90%	60%
WEIGHT/HEIGHT	●		
TRICEPS SKINFOLD	●		
MID UPPER ARM CIRCUMFERENCE	●		
MID UPPER ARM MUSCLE CIRCUMFERENCE	●		
ALBUMIN		●	
CREATININE HEIGHT INDEX	●		
LYMPHOCYTE COUNT			●
IRON BINDING CAPACITY/ TRANSFERRIN			●
CELLULAR IMMUNITY SK/SD			●
OTHER CANDIDA			●
D.N.C.B.			●

Figure 12.7
Assessment of Kwashiorkor Malnutrition

5.1

Enteral Diet

There are numerous special diets for a variety of diseases. Some are low in protein for chronic renal failure; others, low in fat for malabsorptive states; others have fixed caloric counts for diabetics. In the particular context of malnutrition, we are concerned with ways of improving caloric and nitrogen intake.

[1]Keofeed, Stomach Tube, Hedeco, Health Development Corp., 2411 Pulgas Ave., Palo Alto, California 94303; Dobbhoff, Enteric Feeding Tube, Biosearch Medical Products, Inc., Raritan, New Jersey.

We will now consider four varieties of special diets: (1) defined formula diets, (2) low residue diets, (3) feeding modules, and (4) supplements.

I. Defined Formula Diets

These diets are essentially crystalline amino acids or protein hydrolysates, requiring little or no digestion by the alimentary tract. They have no bulk or fiber content. They are of much value when maldigestion occurs, but are unpalatable and poorly tolerated by mouth. Further, they are very active osmotically, and great care must be taken to introduce their use slowly. In order to prevent diarrhea, defined formula diets must be administered at half-strength and then gradually built up to full strength. Examples of defined formula diets (DFD) include Vivonex,[2] Vivonex HN,[3] Flexical,[4] and Aminaid.[5] The last one has no free electrolytes and is therefore especially useful when one needs to reduce the amount of sodium or potassium in the prescription. Most of these elemental diets, however, are expensive and are best used for only a short time.

II. Low-Residue Diet (Meal Replacement)

Low-residue regimens can be taken by mouth or via a nasogastric or jejunal feeding tube. They have a high calorie, high nitrogen content, consisting of caseinated protein with carbohydrate and some fat, and they need some digestion. Most of these products are acidic and contain electrolytes—factors to be remembered in the patient with renal or hepatic failure. Such products are complete foods, contain added vitamins, and are better tolerated than elemental diets by mouth as they are not merely amino acids. They are, however, active osmotically and must be introduced at half-strength and gradually increased in strength and volume over 2 to 4 days. It is advantageous to give the feeding continuously if the tube is in place, thus spreading the osmotic load over 24 hours. Examples of such formula diets include Ensure,[6] Isocal,[7] Nutri-1000,[8] and Precision Low Residue (PLR).[9]

Portagen[10] is of particular interest as it is low in long-chain fats and high in medium-chain triglycerides (MCT), which are better absorbed in maldigestive states than are long-chain fats.

[2]Vivonex, Eaton Laboratories, Pharmaceutical Co., Division of Norwich, 17 Eaton Ave., Norwich, New York 13815.

[3]Vivonex HN, Ibid.

[4]Flexical, Mead Johnson Labs, Mead Johnson and Company, 2404 Pennsylvania Street, Evansville, Indiana 47721.

[5]Aminaid, McGaw Labs, 2525 McGaw Ave., P. O. Box 11887, Santa Ana, California 92711.

[6]Ensure, Ross Laboratories, 625 Cleveland Avenue, Columbus, Ohio 43216.

[7]Isocal, Mead Johnson Labs, Mead Johnson and Company, 2404 Pennsylvania Street, Evansville, Indiana 47721.

[8]Nutri-1000, Syntex Labs, Inc., Stanford Industrial Park, Palo Alto, California 94304.

[9]Precision Low Residue, Doyle Pharmaceutical Co., 5320 West 23 Street, Minneapolis, Minnesota 55416.

[10]Portagen, Mead Johnson, Mead Johnson and Company, 2404 Pennsylvania Street, Evansville, Indiana 47721.

173

III. Feeding Modules

The proper and, in many ways, ideal approach to effective nutritional prescriptions requires the facility to adjust nutrient concentrations in the patient's formula. Feeding modules refer to concentrated sources of a single nutrient, given as an extra to top off defined formulas or ordinary diet. Examples are: (1) EMF (enzymatic modular formula)[11] which contains 15 gm protein per 30 cc, (2) Polycose,[12] a pure carbohydrate source, and (3) Lipomul,[13] which is a fat product. These feeding modules allow one to modify the basic compounded nutrition prescription according to the requirements of the individual patient.

IV. Food Supplements

These are foods rich in protein, with an average protein content of 23%, e.g., Sustacal,[14] Meritene,[15] or Lanolac.[16] Such supplements have a nitrogen:calorie (N:C) ratio of about 1:80, and are thus useful as high protein foods, well tolerated by mouth.

Again, elemental and defined formula diets are very active osmotically, so that care must be taken not to induce dehydration. Also, hyperosmolar non-ketotic coma is a risk with diabetics on high-carbohydrate diets.

Further practicalities include serving these diets chilled (especially elemental and defined-formula diets) if given by mouth to make them more palatable. Head-up positioning for bed-ridden patients on nasogastric tube feeds reduces the risk of aspiration. Ideally, single formula offerings should be limited to 4 ounces every 1 to 2 hours rather than to the usual 8 to 12 ounce prescription, which may overwhelm anorectic patients. For these patients, it is best to replace regular food with a meal substitute formula concentrated to greater than 1.5 kcal/cc, which is palatable to an intake of 32 to 48 ounces per day.

When voluntary daily intake of formula and food is less than 1000 to 1200 kcals, tube feeding at night, using a small silicone rubber feeding tube, is necessary. One can also supplement oral with parenteral feeding. Oral feeding is a far better practice in terms of both effect and cost than putting the patient on nothing by mouth (n.p.o.) and adopting total parenteral nutrition.

Finally, it is important to realize that provision of 100 grams of protein (1½ to 2 grams per kilo per day) is the first goal since protein is the essential nutrient in host defense. By using

---

[11]EMF, Control Drug Co., Market Street, Port Reading, New Jersey.

[12]Polycose, Ross Labs., Div. of Abbott Labs., 625 Cleveland Ave., Columbus, Ohio 43216.

[13]Lipomul, Upjohn, 7171 Pottery Lane, Kalamazoo, Michigan 49002.

[14]Sustacal, Mead Johnson, Mead Johnson and Company, 2404 Pennsylvania Street, Evansville, Indiana 47721.

[15]Meritene, Doyle Pharmaceutical Company, 5320 West 23 Street, Minneapolis, Minnesota 55416.

[16]Lanolac, Mead Johnson, Mead Johnson and Company, 2404 Pennsylvania Street, Evansville, Indiana 47721.

174

liquid protein containing 15 grams per fluid ounce such modular feeding three times a day can provide useful extra protein.

5.2

Parenteral Nutrition

Parenteral hyperalimentation uses a large bore central vein for the purpose of providing all or most of an individual's nutritional needs. As originally described by Dudrick and colleagues at the University of Pennsylvania, this technique utilized the subclavian vein. A mixture of protein hydrolysate, amino acid, and concentrated dextrose solution, with added minerals and vitamins to provide a complete nutritional profile excluding fat, is infused into the vein. A large central vein is necessary as the dextrose concentrations needed to provide adequate calories are hypertonic and would quickly thrombose any peripheral vein. Most often the subclavian vein is used for access, approached by direct infraclavicular puncture (Figure 12.8) so that the tip of the catheter is advanced to lie in the superior vena cava. This method must be explained carefully to the patient, as it is performed under local anesthesia, requiring the patient to be head-down and to have his or her face partially covered by sterile towels during the procedure. The nurse plays a major role in reassuring the patient, in explaining the undertaking, and in following the patient over the course of total parenteral nutrition.

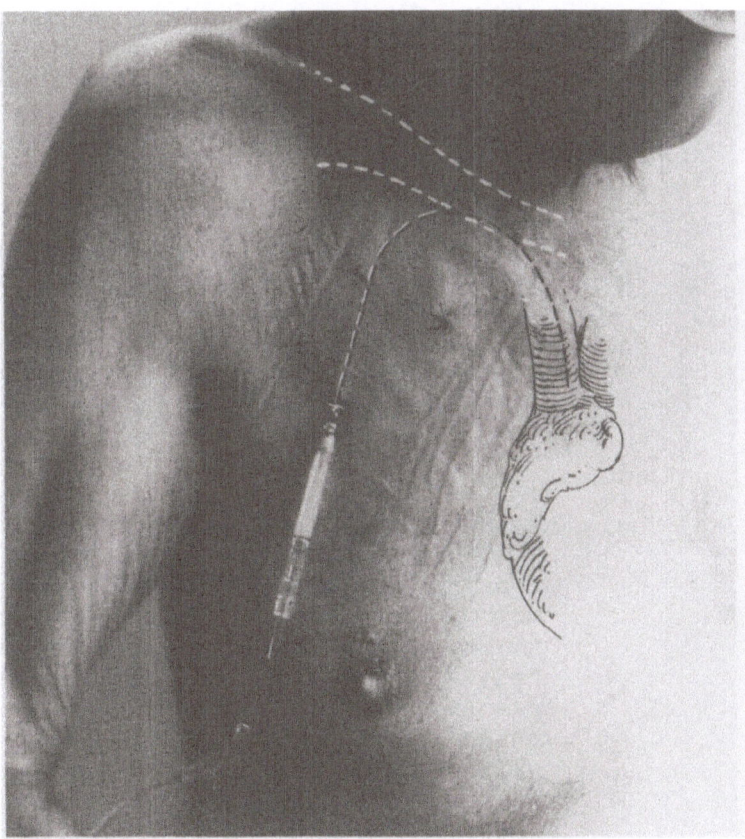

Figure 12.8
Anatomy of subclavian venous catheterization

175

Subclavian catheterization has the advantage of an exit site on the chest, away from any flexures—as movement would constitute a hazard—and away from any orifices so that infection risk is minimized. The patient is thus free to use his or her arms without encumbrance, a distinct advantage with long term hyperalimentation.

Subclavian catheterization is not without risk: a strict protocol and aseptic technique must be carried out during insertion of the catheter. A chest x-ray must be obtained before infusion of any hypertonic solution. Complications may include: pneumothorax due to puncture of the apex of the lung (1% incidence for experienced operators), and arterial puncture. Insertion of the line into the subclavian vein and advancement occasionally result in the entrance of the catheter tip into the internal jugular or even the external jugular vein, necessitating repositioning of the catheter.

The main problems, once the catheter is in place, are related to infection. The catheter is a foreign body containing a highly nutrient solution which is ideal for bacterial growth. Strict precautions must be observed in order to reduce catheter-related sepsis to an acceptable level, which should be under 4%. These measures include:

(1) Strict asepsis in catheter introduction
(2) Preparation of all solutions and additives under a laminar-flow hood
(3) Regular inspection of the catheter-entry site and redressing with antimicrobial ointment every 48 hours
(4) Restrictive use of the hyperalimentation line for that purpose alone. A common cause of sepsis is violation of the line in order to give added medicines, blood, and blood products
(5) Complete change of intravenous tubing every 24 hours

The nurse bears a particular responsibility in these areas and, with careful management, can maintain a central line free of infection for many weeks.

The reduced immunocompetence of malignancy contributes markedly to the threat of infection. For patients on chemotherapy the risk is compounded; precautions must be applied rigidly.

For those patients who receive radiotherapy to the upper chest for breast carcinoma or mediastinal nodes in leukemia states, the exit site of the catheter should be tunnelled subcutaneously well away from the radiotherapy field. Similarly, for patients with tracheostomy or neck wounds, particular care must be taken to keep the skin and dressing around the catheter free of infection. While catheter-related sepsis is a major threat, it is easily treated by catheter removal and replacement in 24 to 48 hours or, alternatively, it may be changed over a wire. Antibiotics are sometimes required.

Three basic solutions are available for total parenteral nutrition (TPN): (1) crystalline amino acids (occasionally caseinated protein), (2) dextrose in a 5 to 50% concentration, and (3) fat in the form of Intralipid.[17] The usual hyperalimentation prescription consists of 25% dextrose with about 4% amino acid with added electrolytes, minerals, and vitamins. This mixture will provide 40 grams (G) protein and about 1000 kcal per liter. Thus, most patients require about 2.5 to 3 liters (L) of this mixture per day. The solution should be administered at a fixed rate over each 24 hour period, with use of an infusion pump recommended. Some patients—the older, the severely stressed, and the septic—and known diabetics will need exogenous insulin to combat the large glucose load. It is important to check the patient's blood sugars initially to detect such diabetic tendency on TPN, and thereafter, at frequent intervals, to check the urine for sugar. Other vital management features for TPN are accurate records of intake and output, along with regular weighings on the same scale every day for early detection of fluid overload (i.e., rapid weight gain).

Intravenous fat is very useful when the fluid volume to be given is limited, as it is the most concentrated form of energy providing 9 kcals/cc. It is also valuable in long term hyperalimentation because it prevents essential fatty acid deficiency and reverses the fatty liver which may occur with long term continuous intravenous hyperalimentation (IVH). However, Intralipid cannot be mixed with other solutions, but must be piggybacked into the infusion just before its entry into the catheter dressing. It is also quite expensive.

Commencement of hyperalimentation must be slow. The patient must work up to the required volume and concentration over several days in order to permit pancreatic adaptation and to reduce the risk of sudden fluid overload. Similarly, hyperalimentation should be discontinued gradually while weaning the patient onto oral or tube feeding. A valuable technique in this respect is cyclic hyperalimentation: the patient is fed intravenously for only part of each 24 hour period, usually overnight, and the line is plugged during the day so that he or she is free to walk around and to eat.

5.3   **The Team Approach**

Nutritional assessment and comprehensive integrated cancer therapy involve a collection of medical staff, nurses, dietitians, pharmacists, and laboratory personnel. With a team approach, the various personnel meet to discuss problems, exchange ideas, and achieve a maximum liaison. It is also very important to involve the patient in his or her therapy. A key effort that takes the whole team to accomplish is the prevention of immobilization, disuse atrophy, fatigue, and depression. An adjustable footboard, a trapeze, and a few simple exercises will give patients a role in their own treatment, together with a sense of identity and belonging, so that they know that they are contributing to their own improvement. Exercises help to promote anabolism, to strengthen the patient, and to improve morale.

[17]Intralipid, Cutter Laboratory, 4th & Parker Streets, Berkeley, California 94710.

177

<table>
<tr><td>6.0</td><td>

Nutrition and Surgery in Cancer Patients

Generally, the grossly undernourished cancer patient will do poorly in surgery. This patient will be subject to increased risk of postoperative complications, such as anastomotic leak, wound dehiscence, or infection. The most common cancer in the United States is of the alimentary tract. Since the function of the gut is intimately concerned with ingestion, digestion, and absorption of food, nutritional support is most often required in this area.

</td></tr>
</table>

6.1

Gastrointestinal Cancer

Many of the factors which produce malnutrition before surgery have already been mentioned: cachexia, obstruction, and maldigestion secondary to pancreatic insufficiency. Because the bowel is among the most metabolically active systems of the body, it is not surprising that intestinal changes do occur with starvation and protein depletion. These changes are probably due to reduced cell proliferation leading to reduced absorptive capacity and less enzymatic activity. To the existing nutritional impairment of alimentary tract malignancy is added the insult of surgery. The patient will have a variable period of ileus and postoperative starvation, during which fluid and electrolyte losses are replenished but not calorie and protein stores. Furthermore, particular problems must be considered in the long term with certain radical surgical procedures. It can be seen that nutritional deficits associated with gut surgery and added to pre-existent metabolic problems will necessitate adequate nutritional assessment, planning, and appropriate therapy both before and after surgery.

If the patient is grossly marasmic or kwashiorkoid preoperatively, he or she should be given about 2 weeks of nutritional rehabilitation, preferably via the enteral route. Parenthetically, the parenteral route is quicker and more certain where GI dysfunction is present. Surgical results improve greatly in such patients after a period of hyperalimentation. There is no evidence in humans that the tumor mass grows disproportionately more than the host. This fear has been raised on the basis of animal experiments related to special strains of inbred rats with transplanted tumors. Other studies, however, have documented advantages from intravenous feeding in tumor-bearing animals.

In tumors of the mouth and oropharynx, where postoperative mechanical problems may arise, tube feeding is the choice with a fully functional gut distally. Esophageal lesions are often associated with marked wasting before operation and will benefit from presurgical IV hyperalimentation, which can be continued afterward over the critical period of anastomotic healing. Oral diet is then reinstituted gradually. Similarly, gastric carcinomata may require radical gastrectomy, a major surgical procedure with substantial risk. Complicated by malnutrition, this procedure can lead to prolonged convalescence with increased morbidity and mortality. Here again, a period of preoperative TPN will lead to a smoother course afterward.

The long term problems of total gastrectomy will not be considered in detail here, but include reduced body weight due to less caloric and protein intake (the result of loss of the gastric reservoir), anemia, reduced fat absorption, and dumping syndrome.

Radical surgery for pancreatic and periampullary carcinomas may be complicated by pancreatic fistulae. Healing requires total bowel rest (as do any high intestinal fistulae) to allow a period of medical treatment so that if further surgery is necessary, the patient is in a fit state to withstand the added surgical stress. Accordingly, TPN is the approach in such patients.

Massive small bowel resection of lymphomas may result in the "short bowel syndrome"—a major challenge in nutritional management necessitating permanent intravenous hyperalimentation with an "artificial gut." This management may be carried out at home by the patient. Large bowel cancers are not usually associated with major nutritional problems before surgery, but a complicated postoperative course may result from low anterior resection which leaks, and nutritional support may be required.

Nutrition and Radiotherapy

6.2

Radiotherapy achieves its effect by destroying the most actively dividing cells in this field. These should be the cancer cells. However, normal cells of the gut proliferate quickly, particularly in the small bowel mucosa, and, with the bone marrow, are among the most active of normal tissues. Thus, they are particularly vulnerable to radiation damage resulting in secondary stomatitis when the mouth is involved, or with radiation enteritis if abdominal and pelvic areas have been irradiated.

Soft-tissue tumors of the mouth and pharynx are relatively radiosensitive and are frequently associated with radionecrosis, a cause of sore throat, pain on swallowing, dry mouth and (most significantly) reduced and altered taste sensation, often amounting to "mouth blindness." Physical examination may reveal ulceration, edema, and a pseudomembrane. Weight loss is common, and nutritional support becomes necessary. Such measures should include attention to oral and dental hygiene, while blenderized or liquid diets may overcome the pain of mastication. In some cases tube feeding will be mandatory. TPN may serve as minimized radiation reaction, allowing weight gain. Thoracic radiotherapy often produces difficulty in swallowing severe enough to warrant a feeding tube.

Irradiation of the abdomen may induce increased protein losses from the bowel as well as reduced intake due to enteritis. Incidence of radiation-induced enteritis is not known, but 8% of 384 patients in one study of pelvic irradiation for bladder cancer suffered severe bowel symptoms. Traditional remedies include antiemetics, antispasmodics, antidiarrheal agents, with a bland diet and occasional steroid enemata.

Specific dietary therapy with low-residue, low-fat, lactose-free regimens for children given abdominal irradiation has been useful. Defined-formula diets may well protect against radiation enteropathy, while intravenous hyperalimentation will make it possible for poorly nourished patients to withstand full-dose irradiation.

6.3

Nutrition and Chemotherapy

Chemotherapeutic agents are toxic and commonly produce side effects on the bone marrow and the bowel. The dose-limiting factor is usually the hemopoietic system with reduced resistance to infection, which is compounded by the malnutrition induced by concomitant anorexia, nausea, vomiting, stomatitis, and diarrhea. When intravenous hyperalimentation is contemplated, these effects, induced by chemotherapy, pose a potential hazard of overwhelming bacterial or fungal catheter-related infection while disordering clotting parameters, a problem related to catheter insertion. These risks do not apply to enteral feeding, but it may be impossible or of reduced value because of the side effects of chemotherapy.

Nausea and vomiting are common and appear to be triggered by a central effect of the drugs on the chemoreceptor center. They are not wholly combatted by antiemetics, of which phenothiazines are probably the most effective. While most chemotherapeutic agents induce these symptoms, steroids, busulfan, and vincristine are notably troublefree in this respect. Stomatitis is common, and it is notable that tumors of the bowel do not grow as quickly as normal small bowel mucosa, which explains the sensitivity of the bowel to chemotherapeutic agents. The stomatitis of 5-Fluorouracil given as an infusion is often dose-limiting, and studies in both rat and man have revealed a greater tolerance with fewer toxic effects, allowing a larger dose when adjuvant intravenous hyperalimentation is utilized. Methotrexate and actinomycin-D are also prone to induce stomatitis, which will result in anorexia and "mouth blindness." Vincristine is notable for inducing ileus and constipation in up to 30% of patients, and though there is little documentation of malabsorption in patients receiving chemotherapy, it is likely that the small bowel absorptive mechanism is disturbed. Concern for esophageal erosion during chemotherapy is unwarranted with use of a small 9F silicone feeding catheter. In fact, this technique is often ideal to provide the slow feeding that will not invoke nausea and vomiting.

7.0

Summary

It has become apparent that weight loss and associated fatigue, depression, and intercurrent illnesses are not an obligatory consequence of cancer. It is absolutely essential that every student and provider of cancer care appreciate the role of nutrition in the treatment of malignancies. Future modification of the North American diet can be expected to have some favorable influence on the incidence of many cancers as well as of cardiovascular disease and obesity. Universal consumption of fewer calories, particularly fried meat, fat, sugar, and salt, is a reasonable goal at this time.

But once cancer has progressed to loss of appetite and weight, dietary considerations must be directed to favorable protein and caloric intake by whatever form and route possible. Proper concern for the patient and conscientious care include constant surveillance for signs and symptoms suggesting anorexia and for the evidence of serial weighings of the patient to detect weight loss. Once it reaches significant proportions—such as a 10 pound or more weight loss, lymphopenia, a falling serum albumin, or, most importantly, anergy to delayed hypersensitivity skin antigen testing—vigorous nutritional treatment must be started.

References

Conference on Nutrition and Cancer Therapy, *Cancer Research*, **37**, 2321–2471 (1977).

Blackburn, G. L., and Bistrian, B. R., Nutritional counselling: role of nutrition support services. *In:* Schneider, H. (ed.), *Nutritional Support Medical Practice*. New York, Harper & Row (1976).

Copeland, E. M., MacFayden, B. V., Lanzotti, V. J., and Dudrick, S. J., Nutritional care of the cancer patient. *In:* C. E. Howe (ed.)., *Cancer patient care at M. D. Anderson Hospital and Tumor Institute*, Chicago, Year Book Medical Publishers, (1976).

A. M. A., Defined Formula Diets for medical purposes, A. M. A., Chicago, Illinois (1975).

Fischer, J. E., (ed.), *Total Parenteral Nutrition*. Boston, Little Brown and Company, (1976).

Blackburn, G. L., Bistrian, B. R., Maini, R. S., Schlamm, H. T., and Smith, M. F., Nutritional and Metabolic Assessment of the Hospitalized Patients. *J. of Parenteral and Enteral Nutrition*, **1**, 11–19.

13 Psychodynamics in the Cancer Patient

Chapter 13
Psychodynamics of the Cancer Patient

1.0

Background

The psychological impact of cancer and its treatment is enormous for both the patient and the patient's family. There are many factors which contribute to this impact: disease and treatment variables, the patient's age and socioeconomic group, and the patient's unique psychologic attributes. Each patient has a complex and individual emotional configuration. Two people of the same age, from the same socio-economic group, receiving similar treatments for the same disease may react quite differently to their experiences. Furthermore, just as individuals differ in their reaction patterns, so do families.

This chapter proposes a way for health care teams to understand patients with cancer by learning about them from their experiences. It will be demonstrated that patients' responses to cancer and treatment are logical and can be understood as a systematic process when one becomes better acquainted with the patient and the family.

2.0

Identifying the Patient's Dilemma

Several young people, ages 10 to 17, with osteogenic sarcoma who had been treated with leg amputation and high dose methotrexate with subsequent nausea, vomiting, and hair loss were asked what was the worst part of their treatment. Each reported that hair loss, nausea, and vomiting were more upsetting than limb amputation. Although nausea and vomiting were extremely uncomfortable and amputation was a loss, these patients and their friends could accept amputation better than other side effects. It seems surprising that temporary losses and discomforts were more important than the permanent injury of amputation.

A woman who lost her job through an unfair power play at work during the spring, who underwent mastectomy the following winter, and whose father died suddenly several weeks subsequent to the mastectomy reported that, of all these misfortunes, job loss was the most upsetting. Her reaction seems illogical; job loss is a temporary issue while loss of breast, threat of cancer, and loss of father are all permanent and seemingly more devastating injuries.

How should the health care team deal with reports such as these? Should they be taken lightly or seriously? This question is important; by answering it the health care team must reflect upon patients' reactions to cancer and the treatments they receive.

When these patients revealed that hair loss and job loss made them feel worse about themselves than amputation or mastectomy, they were clearly talking about injuries to their self-esteem. In this chapter, we will analyze these injuries and explore how they are brought about by cancer and external life stresses. We will also consider the verbal and physical behavior of patients and their families as signs of how they are coping with these injuries and the feeling provoked by them.

3.0   **The Central Issue: Self-Esteem**
There are many psychosocial variables that operate in patients' lives in the course of illness. This chapter proposes that self-esteem is the most important variable and that psychosocial is not simply a descriptive term of variables but refers to a process going on in the lives of patients and their families.

Self-esteem is a person's appraisal of his or her own worth. A person strives to maintain adequate self-esteem. One can consider the self as having four parts: 1) body self, 2) interpersonal self, 3) achievement self, and 4) identification self. Identification self generates self-esteem through being affiliated with a person, group, cause, or possession, which the person and usually the world consider worthy. This form of self-esteem includes knowing important people, rooting for winning teams, church, charity, political memberships, and having a fine house, car, or boat. It is the sum or total feeling about one's condition and activity which makes up the total self-esteem. Activities of daily life are thought to generate feelings of satisfaction or self-esteem.

Clearly, the primary impact of cancer is on the body self. At cancer diagnosis the patient is told that there is a disease in his body which is a threat to his life. The body is then often further injured by treatment, bringing disfigurement, pain, and discomfort. The cancer patient also suffers interpersonal injuries when intimates and friends draw away. He or she often withdraws into isolation, thus aggravating the interpersonal self-injury. Cancer often means a reduction in daily activities either at home, on the job, or at leisure, caused by physical disability, loss of interest, or both. Such curtailed activity reduces satisfaction gained from achievements. Cancer can also jeopardize sources of identification self-esteem. For the religious person, the diagnosis of cancer can put a strain on life-long beliefs and values. If the patient believes that God controls the world, feelings of guilt and punishment may arise as well as hostility towards God. The patient may ask: "Why me?" Faith is difficult to maintain with a progressively worsening course. Cancer can also put a financial strain on the patient whereupon material possessions become a burden rather than a source of pride.

It is now apparent that from a psychodynamic perspective cancer can aggravate if not undermine one's relationship to the sources which generate everyday self-esteem. The problem of adjustment in cancer is often how one can continue to generate positive feelings about the self, despite the injuries or threats of injury which cancer can bring to the body, one's interpersonal relations, one's daily achievements, and one's affiliations. Thus, cancer is not only an injury, but an insult. The psychological response to reduced self-esteem prompts negative feelings, such as distress and anger. These feelings are sometimes felt by the cancer patient on the conscious level, but at other times are so overwhelming and unacceptable that they are relegated to an unconscious level. Patients feel that it is outrageous to get cancer, and disfiguring treatment only compounds their injury.

4.0   **Methods of Coping**

When we speak of coping with cancer, we mean the process of how people maintain self-esteem under the stress of cancer and its treatment. Coping can mean sustaining normal activity, keeping going, not letting cancer interrupt those activities which give one a sense of satisfaction and worth. Coping also means compensatory activity, however, which tries to repair or replace self-esteem injuries. One frequently observes cancer patients increasing their energies, achievements, and goals after diagnosis in ways that are clearly compensating for the losses they have felt. Maladaptive coping is behavior which gives up maintaining self-esteem or even leads to further self-esteem loss. What is remarkable to the observer without cancer, however, is how well cancer patients cope with this injury and deal with stress and distress.

Emotional defenses are an important unconscious aspect of the coping process. A defense is defined as a "mechanism of adaptation selected unconsciously, operated automatically, to manage anxiety, impulses, hostility, resentments, and frustration" (Noyes and Kolb, *Modern Clinical Psychiatry*). Originally described in the field of psychoanalysis, the definition and categorization of defenses is still somewhat arbitrary, but familiar examples are denial, distortion, projection, rationalization, reaction formation, and sublimation. Defenses are by no means exclusive processes of the cancer patient. They are a central part of everyday life as important as the body's physiological defenses. However, in the diagnosis of cancer, they are relied on more strongly and are more easily observed. The strength of the emotional defensive system as an adaptive resource is one reason why cancer patients do so well.

4.1   **Denial**
A measure of adaptive healthy denial is necessary for patients and families to be able to function with cancer. Denial allows the cancer patient to continue normal daily activity without constant preoccupation with the full grim reality of illness. If the cancer patient was continuously in touch with the negative feelings which the stresses of cancer incite, he or she would be unable even to function from day to day. To the person without cancer, this denial sometimes seems inappropriate, indicating that the patient is failing to realize the importance of his or her situation. Cancer professionals, however, have learned to welcome denial as a normal essential adaptive process which allows people to go on living without over-preoccupation with illness. In many cases the greater the stress, the stronger the denial. A 46 year old woman, 18 months post-mastectomy, was able to discuss feelings of lost sexual attractiveness, anger about her operation, lopsidedness, and feeling a little like a freak. However, several months later, when interviewed in the hospital for treatment of recurrent disease, she completely denied negative feelings about her mastectomy. She was matter-of-fact about the location of her new lesions, and in a blasé manner reported that the cancer was spreading. Her use of denial had increased commensurately with severe multiple stresses and with her need not to be overwhelmed by her negative feelings. Thus, defensive processes are not rigid mechanisms, but are flexible, capable of shifting in intensity to fit one's needs.

4.2   **Projection**
The defense of projection—that is, attributing to another what one is feeling about oneself—is another way cancer patients deal with feelings of reduced self-esteem. Instead of expressing feelings of their own inadequacy, of worthlessness, and of a lowered self-image, they find fault with their environment from care of food to the hospital ward and the doctors.

4.3   **Distortion**
The defense of distortion is the mechanism of altering reality to suit one's needs and to protect oneself against distress. A physician proposes to a 42 year old woman that the best way to treat her colon tumor is by chemotherapy and radiation. She

remarks that she's glad that she has a tumor and not cancer because the tumor is benign and cancer will kill you. Distortion is usually evident when there are continual misunderstandings despite more than adequate explanation. The physician should understand that this is a defensive process and not a result of inadequate communication. The prudent physician checks from time to time to ensure that the families of patients who engage in distortion fully understand treatment plans and their implications.

4.4   **Other Defense Mechanisms**
The defenses mentioned above protect the patient emotionally by obscuring reality to a certain degree. Other, more flexible defenses are used when the emotional injury is not as great or when unacceptable negative feelings are not as strong. Healthier patients are generally able to handle their injuries without major use of these primitive defenses and rely on defenses like rationalization, reaction formation, and sublimation to handle stresses and injuries.

A 33 year old male undergoing radiation therapy to treat his sarcoma reported he was lucky compared to other patients undergoing treatment concurrently because they were much younger and sicker. This rationalization or "putting things into perspective" is not distorting reality but is nevertheless a defensive process to protect the patient from his own despondent situation. In sublimation, negative feelings are redirected and utilized towards a constructive or satisfying result. Feelings are acknowledged but vented in ways that are neither self-destructive nor unpleasant to others. Art is a classical example of a positive redirection of unconscious feelings. In cancer, sublimation is evident when patients harness distressing energy to grow emotionally. They write books, become involved in cancer societies, devote new energy to bringing their families closer together, or devote themselves to other needs of an altruistic nature. This reponse is creative as well as adaptive; patients recognize their need for emotional outlets and are mature enough to devise their own therapy.

All of these examples of patients' defenses suggest that their responses to cancer are not irrational but logical. We can understand why the young patients described earlier found hair loss a greater stress on their self-esteem than amputation. Baldness made them feel closer to death and encouraged peer discomfort and rejection. We sympathize with the woman who found job loss more upsetting than her mastectomy and her father's death. Her job obviously generated her daily sense of self-esteem. Finally, as we have seen, a degree of healthy denial is also logical as it prevents patients from fully succumbing to a grim daily awareness of cancer.

5.0   **The Role of the "Closest Person Emotionally" (CPE)**
The impact of cancer diagnosis on the patient is sometimes more clearly evident after talking with the spouse or other

closest person emotionally, the CPE. Characteristically, a portion of the hostility generated by cancer is projected towards the CPE be it spouse, parent, or child. The CPE may discover that he or she is now being blamed for things that go wrong, is being made to feel inadequate, or is often put in a powerless position where he or she is "damned if I do and damned if I don't." A mild example of this phenomenon is a 34 year old man with lymphoma who continues to tell his wife that he can't cope with daily life. She makes numerous suggestions to help him but none appear to work for him. It turns out that part of his actual method of coping is to state that he can't cope and then see to it, without realizing it, that all his wife's efforts fail. If she stops trying, she risks his saying that she doesn't care about him anymore; if she continues, she continues to fail and to be frustrated. In this way, curiously, he is able to arouse in her a feeling of helplessness. In an experiential way, this tactic allows him to share with her uncomfortable feelings of helplessness which he is having secondary to his illness.

A more extreme example is a 56 year old woman with breast cancer, under several stresses, who says that her husband is unfaithful to her. This is a delusion which is helping her to handle the impact of the stresses. She resists all efforts by him to convince her that he is faithful. If he continues to try to convince her, she becomes angry and increases her accusations. If he stops trying to convince her, she interprets his silence as his frank admission of guilt and also attacks him for that.

In short, the person closest to the patient is made to feel helpless at a time of increased stress. It is important to identify this person, to understand this process, and to help this person cope with the patient. In evaluating the impact of illness and treatment, the patient's style of generating self-esteem, of handling hostility, and of coping must be considered. This evaluation can be done in a reasonably systematic way and can be conducted along the same lines as one evaluates other body systems.

6.0   **Psychosocial Evaluation of the Cancer Patient**

An essential ingredient in the emotional care of patients is the relationship between the caregivers and the patient and the patient's family. This is self-evident and traditional. Each physician or caregiver forms the relationship in which he or she is the most comfortable and which is most characteristic of his or her style. What matters is that there is a sense of trust and confidence, for the patient looks to the physician or caregiver for guidance and is willing to accept limits when they seem necessary.

The physician makes an emotional evaluation in a commonsense way when he or she has first contact with the patient. The physician intuitively compares later observations with the baseline interpreted in the light of disease course, treatment course, and outside life events. The major concern is how the patient and family deal with the stresses of illness and treatment, particularly when these stresses are severe.

6.1

Identification of Risk Factors

The initial psychosocial evaluation should take into account: 1) an assessment of the number of stresses or risk factors being put on the patient at that time, and 2) an evaluation of the patient's capacity to handle stress. Risk factors can include not only the side effects brought on by treatment and other difficulties related to illness but concurrent life stresses unrelated to illness, such as recent loss of a family member, financial difficulties, business reversal, and children in trouble. An assessment of the patient's capacity to handle stress includes such factors as previous psychiatric illness (especially depression), chronic difficulty in handling hostility, a weak social support system, and long standing low self-esteem.

Risk factors combine either to weaken the patient's capacity to cope under stress or to increase the existing stress and further injure self-esteem. A 48 year old woman had bilateral mastectomies. Shortly after she finished adjuvant chemotherapy, her physician moved away and she was forced to switch doctors. The whole experience led to emotional decompensation and she required psychiatric hospitalization. Psychiatric treatment revealed that her increased susceptibility to the anxiety at the end of her medical treatment and the loss of a highly prized physician were related to unresolved feelings secondary to losing her mother at the age of five through divorce and then living with her father. This susceptibility, added to the chronic injury of bilateral breast loss, resulted in decompensation.

In general, one or a combination of these risk factors impairs a patient's capacity to cope optimally with the stress of cancer and treatment and increases the likelihood of a pathological emotional reaction at a stressful time.

6.2

Assessment of Self-Esteem Quotient

An effective psychosocial evaluation will attempt to objectively gain a picture of each person's self-esteem. A useful strategy is to have the patient describe his or her typical day. With daily activities as a background, questions then can be directed towards the four self-esteem areas of body self, interpersonal self, achievement self, and identification self. A few questions should be directed towards how the patient has coped with previous self-esteem injuries, such as previous surgeries, loss of loved ones, or job loss. Adaptive processes are generally consistent over time and it can be expected that patients will cope with cancer as they have coped with other severe difficulties in their lives.

Because the CPE is the major source of support for the patient as well as the primary emotional outlet for negative feelings, it is a good idea to include the CPE in the psychosocial evaluation. Sometimes by providing support to the CPE, the patient receives the best possible emotional support in an indirect manner. Once communication with the CPE is established, the CPE becomes extremely valuable, serving as a sentinel or monitor of the patient's behavior. Discrepancies in the patient's and CPE's report of how the patient is coping can serve as "red flags" that all is not as well as the patient would like the physician to believe. Patients sometimes resent inclusion of the CPE in their discussions with doctors and feel that it is an intrusion into a private personal relationship.

In such cases, a judicious mix of some private time with the patient and some joint time with patient and CPE is recommended with the firm, kindly assurance that the patient and CPE are part of the emotional process, and both must be included in emotional evaluation and treatment. It is also important to monitor the CPE. A general knowledge of the way in which CPEs function emotionally within their self-esteem systems is very helpful in assessing whether, at one point or another, they are managing their own lives successfully or whether they are decompensating in response to the stress of the patient's disease and treatment.

7.0

Specific Patient Needs and Management
For almost any cancer patient, and frequently for family members, there are some reasonably predictable emotional issues which generally have to be addressed. While each person reacts in an individual way, there are general patterns which most people follow. There are issues the physician can discuss with patients from time to time or with patients and families. The more they can be identified and talked about reasonably, the more likely the patient will be able to handle the emotional burden of illness and treatment effectively.

7.1

Career Planning
Each person has a career or life plan which he or she is striving to fulfill. The cancer patient often must telescope his or her career for the next 20 or 30 years to a 2 or 3 year timetable. If a 28 year old mother with a 6 month old baby has learned she has lymphoma, she must redirect her hopes of seeing her child be married and produce grandchildren to seeing the child learn to speak, walk, and go to school. Hopes must be therefore reshuffled into realizable expectations. Although cancer is a tragedy causing sadness and heartache, healthy coping must at some point acknowledge that while dying is a major problem, another problem is what the patient is going to do with the remainder of his or her life. Goals, hopes, and achievements can still be maintained and satisfied, but must be adjusted to a reduced life span and often to a reduced activity level.

Emotional Support

Self-esteem is what needs support and it is often supported extremely well by intuitive caregivers. Is support being sympathetic? Is it being comforting? Is it being firm? Is it leaning over backwards to do anything to help? The answer, of course, depends upon the situation. The best support is to diagnose what gives the patient self-esteem—be it work, a relationship with a loved one, parents, or autonomy—and then to encourage or facilitate these sources of support. Being overly sympathetic or bending over backwards to answer every request often has the reverse effect of reinforcing in the patient the idea that cancer automatically puts one in an inferior or dependent role. Sometimes if the ability to continue functioning as previously is not even made an issue, that is the best support possible.

Vicissitudes of Daily Life

Cancer and treatment are not only an obvious additional stress for patients but they lower a person's threshold for handling the wear and tear of daily life. Patients are more likely to be very irritable after being cut off in traffic or furious when the switchboard operator cuts off a phone call. Patients experience more extreme reactions than they would under normal circumstances. For example, a 37 year old man with lymphoma found himself infuriated at his 5 year old child who wouldn't eat his string beans at dinner. A 56 year old woman with breast cancer fell apart when her youngest son went off to college.

Helplessness

A feeling of helplessness—the feeling that cancer has developed and that it will probably prevail—is commonplace in cancer patients and in their families. Helplessness is an unwelcome, frustrating, and often infuriating feeling. It is particularly strong at times of initial diagnosis, of relapse or recurrence, or of serious outside stress or loss. Patients are frequently not aware of how helpless they are feeling. Coping mechanisms mercifully protect them from the full intensity of this disquieting awareness.

It is at these times that the patient is likely to put his or her CPE into a helpless position without realizing it, as discussed in section 5.0. The patient's feelings of helplessness and frustration, too great to be allowed into consciousness, are curiously transferred to the CPE. The CPE ends up feeling helpless about something (the subject matter is unimportant). This transferal of anxiety is often considered a "putdown." This description is accurate as the CPE's self-esteem is assaulted and the CPE does feel put down. While this maneuver relieves the patient, it can be disruptive to the relationship with the CPE. The CPE, feeling helpless and angry, may withdraw at the very time when he or she is needed most. Furthermore, the patient may feel guilty for provoking this reaction.

The helpless and often hostile interchange between patients and CPEs is quite frequent. Despite its frequency, most patients and CPEs never even consider the possibility of discussing their discomfort with the doctor. Most would regard such a confession as an admission of weakness and as a source of embarrassment. Physicians should explain to patients and CPEs that hostility and helplessness are expected and normal coping reactions and should inquire about such tensions at each visit. From an acknowledgement and expression of negative feelings toward each other, patients and CPEs can, under the guidance of a physician or other caregiver, explore issues of anger, guilt, discouragement, depression, and fears of dying.

7.5

Guilt

Guilt in the cancer patient and CPE often involves the process of feeling some responsibility for the illness. Offers to ease guilt are often met with considerable resistance, which should be respected. Guilt can give patients a sense of control over their plight without which they would feel at the mercy of an irrational force. Also, it has been noted that when CPEs relieve their guilt, they often feel extremely hostile toward the patient for becoming sick and being a financial and emotional burden. As with other coping responses, guilt, in itself considered an uncomfortable emotion, can serve a reasonably useful defensive purpose.

8.0

Psychological Issues in the Patient-Health Care Team Interaction

The relationship between patient, physician, and health care team is an essential and traditionally well understood part of the treatment process. Mutual trust and respect must be part of this process. With cancer patients, family involvement, as discussed, is also a very important ingredient. The severe self-esteem injuries involved with the disease itself and the treatments which the physician renders frequently stir up strong negative feelings toward the physician in both patients and family members. These feelings require careful attention by both the health care team and the patient to keep mutual trust and respect alive. Sometimes the health care team is split into good, caring people and bad, uncaring people in the eyes of the patient and family. It is incumbent on those caregivers considered good and caring to diagnose this splitting. As gratified as they may be to be well-liked, they should help the patients and families realize that another doctor probably has similar virtues or strengths. It is natural for patients to have these feelings as a way of dealing with the stress of illness.

The defense of reaction formation frequently occurs with patients and families. This mechanism consists of reversing how one feels. Typically, strong inner negative feelings can be turned into strong, often exaggerated, positive ones. While pos-

194

itive, fond, loyal, grateful feelings are desirable and very help-ful, deification or putting the physician on a pedestal often has this defensive component and can increase the distance between the physician and his patient.

8.1

Treatment Compliance

Few patients withdraw from treatment regimens. Once they have agreed to treatment, they tend to stay, for they know the alternative. If patients begin to miss appointments, show up late, or express desires to stop treatment, physicians should look to see if there isn't a major emotional distress. A 22 year old patient with Hodgkin's disease began arriving late to his appointments and missing other appointments altogether. Upon investigation, it was learned that his parents were divorced, he was living with a marginally compensated mother, and had been very dependent on his physician. The distress arose when he was transferred to the care of another physician; he expressed his anxiety in his failure to show up for appointments. A 36 year old woman with bilateral mastectomies started and then refused adjuvant chemotherapy treatment. It was determined that her husband had been depressed for several years, and her refusal was a request for him to resume the decision-making process—a role which he had performed in the past before he was depressed. After he received short-term psychotherapy, she resumed treatment.

8.2

Entitlement

In attempts to achieve compensation for their emotional loss, some cancer patients become infused with a sense of entitlement. It is a common philosophy in our society for people to get compensated for accidents, loss, or misfortune. In cancer patients, this sense of entitlement can be expressed by demanding special treatment, wanting unlimited access to physicians' time, and making other people make plans around their wishes and desires. Cancer physicians state that it takes many times longer to treat a cancer patient than other medical patients. When a 17 year old female with sarcoma who had been acting childish, petulant, and demanding was pressed for why she was acting in such a difficult manner she was able to say that she felt since she had cancer and was going to die she had the right to do anything or be any way she wanted to be.

In these situations it is best to give patients permission to express their furiousness openly, but then set limits about how they must act in treatment. This patient was able to become more manageable after firm limits were set and she was reminded that it was difficult enough for her parents to have their 17 year old daughter dying of cancer without her acting like a 5 year old child. If limits can be set on a sense of entitlement, more adaptive outlets for hostility may become established.

Hostility

It can be seen that many problems of the cancer patient revolve around issues of hostility. Hostility is a natural response to loss of self-esteem. This is evident not only in cancer, but in war, social prejudice, the sports arena, and unemployment. Hostility is also a deep-rooted and instinctual emotion, not always expressed at a conscious level. It can be especially destructive in its more subtle and hidden expressions. It is best to deal with hostility in ways that are neither destructive to the patient nor to the caretakers giving treatment. An example of a subtle expression of hostility occurred when a young sarcoma patient, who was outwardly agreeable and cooperative, had friends and family who were in frequent conflict with nurses and doctors treating her. It was discovered that the patient was casually remarking to her friends and family that she did not mind her lunch being an hour late and cold, that the nurse had not really meant to infiltrate her intravenous line, and that the nurses must work so hard because they never answered her calls. This passive aggressive hostility was not manifested in her behavior but in provoking others to express it for her.

When discussing the cancer patient's social support system, a scapegoat often appears. Somebody has let the patient down, is of no help whatsoever, and is actually a hindrance to his or her daily life. Sometimes the scapegoat is conveniently distant. For example, one woman expressed extreme hatred for her daughter's school principal; another patient demonstrated extreme hatred for the Germans. However, at other times more familiar scapegoats are found. Often the doctor is scapegoated for failing to cure the patient. A golden rule for caretakers is to be able to accept the hostility but not the blame, and to remind the patient that treatment is aimed towards a cure but that there is no guarantee.

One 56 year old cancer patient would periodically begin complaining about her "useless" husband. It was found over time that these complaints were her announcement that she was again preoccupied with dying, terrified of the prospect, and needed to discuss her fears with someone.

Death Perception, Depression, and Anxiety

The literature on death and dying is increasing. Dying with dignity or dying gracefully is dying with the optimal preservation of emotional equilibrium. It is felt that there is not a right or proper way to die as such, but that dying, like living, is an individual process. People will generally die according to the same patterns as they have lived. The gregarious and outgoing person is more likely to have friends and relatives around him at death than the generally isolated, more solitary person.

As death nears, there is a defensive shift to cope with increased stress and fear. Patients undergo to some extent the sequence of denial, anger, bargaining, depression, and acceptance described by Elizabeth Kubler-Ross in her acclaimed book *On Death and Dying.* If acceptance is not forthcoming, often primitive defenses of distortion, projection, and denial are heavily relied on to protect the patient against a very frightening prognosis. Their defenses require considerable emotional energy to maintain and will at times break down. At these times the patient will have outbursts of anger or tears. More adaptive patients can continue to mobilize interpersonal, achievement, and identification self-esteem even though the body self is near demise. At this time religious beliefs often serve as a valuable emotional balance for maintaining feelings of purpose, meaning, continuity, and feeling loved.

Recent studies indicate that there is less depression in cancer patients than was commonly supposed. Depression, however, is often an appropriate response when an additional stress of illness or treatment is encountered. These situations can occur when recurrence of disease is diagnosed, when therapies to control metastatic disease are noted to be ineffective, or when outside stresses impose upon the burden of disease and treatment.

There are also several situations which provoke anxiety, such as awaiting biopsy for a possible malignancy, awaiting results from a metastatic series, or from a bone scan. Absence of anxiety can mean massive use of defenses and can be a distress signal and maladaptive coping device. Delay in seeking treatment for a suspected malignant lesion often results from an inappropriate lack of anxiety. Some patients or families experience an inappropriate or a chronically exacerbated state of anxiety with a resulting sense of discomfort. Reassurance, often limit-setting, hearing out reasonably sounding anxieties, and anti-anxiety medication can all help lessen discomfort and control anxiety.

10.0   **Mental Status Changes in Cancer Patients**
Psychotic reactions occurring in the midst of accumulating stresses on a perhaps already emotionally compromised patient are not uncommon. Such responses, at least in previously normal people, are generally situational and can be reasonably counteracted by a combination of anti-depressant and anti-psychotic medications, increased support, active psychosocial intervention, and at times psychiatric hospitalization. It is important, indeed essential, to rule out a possible organic basis of psychosis before confirming the diagnosis as a response to increased stress by an emotionally compromised host.

10.1   **Response in the Previously Mentally Ill**
A patient with a diagnosis of psychiatric illness can pose special management problems. The decompensation of a 36 year old schizophrenic female was precipitated by her feelings of rejection by her specialist when returned to her family doctor.

197

A manic depressive breast cancer patient required psychiatric hospitalization after the diagnosis of recurrent disease. A paranoid schizophrenic 56 year old man with lymphoma was stable medically, but when his 19 year old daughter experienced difficulties in college and adopted a somewhat different life style, he became psychotic, struck his wife, and required intensive intervention.

10.2 Psychosis in the Previously Healthy Person

A 56 year old nurse with lung cancer had been able to maintain full activity at work and at home until diagnosed with advanced disease. When she was no longer able to run her household and became dependent on her family for care, she developed the delusion that her 4 year old grandson's life was in imminent danger. She had to be physically restrained from trying to protect him. A woman with rectal carcinoma, who had no history of depression, developed a psychotic depression following diagnosis of advanced disease, rapidly developing paralysis in her right leg with subsequent inability to work. She had raised a healthy family and had worked full time as an administrative assistant for many years. In normally healthy patients, the psychotic process was a temporary "time out" to handle overwhelming stresses. Denial, distortion and projection were used to protect the patients from the multiple self-esteem injuries of recurrent disease and job loss.

Both women must have derived a great deal of self-esteem from their work performance. The concurrent injuries of advancing disease and job loss necessitated their use of strong, "psychotic" defenses to alter the overwhelming reality. In normally healthy patients, such responses should be diagnosed as psychological decompensation requiring acute interventions. These patients usually respond well to a combination of psychosocial intervention for treating their self-esteem issues and a drug regimen of psychotropic and anti-depressant medication.

10.3 Acute Brain Syndromes

Changes in mental status can occur secondary to toxic and therefore stressful organic states as well as to stressful psychological circumstances. Delirium is a symptom complex which usually develops acutely, runs a limited course, and shows fluctuations in degree of intensity and level of consciousness. It is characterized by a dreamlike quality, restlessness, often stupor, confusion, and sometimes hallucinations. It is often more intense at night. This condition is usually secondary to conditions which impair function including fever, toxic states, circulatory insufficiency, and metabolic imbalance (see Chapter 11).

One important type of acute brain syndrome is that induced by drugs or common medications. Cortisone induced changes can increase the sense of euphoria, and can prompt hypomania or depression. Delusions, grandiose thoughts, and persecutory ideas often accompany changes in affect. Motor changes can range from hyper to hypo activity. The pattern response depends upon previous personality structure. Other drugs which affect mentation include the phenothiazines, narcotics, sedatives, hypnotics, and mood altering drugs, such as tranquilizers or relaxants.

11.0   **Summary**

The psychodynamics of the cancer patient involve a variety of coping strategies to maintain and restore injured self-esteem. As we have seen, this process is the core issue in the psychotherapeutic support of the cancer patient. Hostility is the common response to self-esteem loss and is expressed in a multitude of behaviors not always on a conscious level. However, if permitted natural expression directed towards constructive outcomes, most people will cope remarkably well.

Traditional patient care including being available, open, and supportive makes for common sense and effective treatment. An open and trusting relationship with the physician can maintain hopes, provide an opportunity to discuss emotional issues and can thus be an enormous self-esteem resource for the patient. The physician can serve as educator in teaching patients and families what to expect emotionally as well as physically in cancer. A psychosocial evaluation and intervention should emphasize that continuing to do the things which are important, satisfying, and enjoyable are central to a healthy coping process and to physical health as well. Permission by the physician to at times feel angry, depressed, and anxious can be enormously helpful in preventing the patient from thinking that he or she is decompensating emotionally. When these feelings occur, anticipation is also a defense and is useful in helping to maintain control when tears or outrage are loosened. Permission to relate emotional symptoms to the physician will also allow the physician to monitor disturbing emotions as well as physical distress throughout the disease course.

CPEs can be educated in their role as a monitor and possible hostility bearer. It should be emphasized that emotional symptoms, such as loss of sexual interest, social withdrawal, and other self-destructive behaviors, are as important as physical symptoms. They should be brought up to the physician just as one would report shortness of breath or recurrent pain.

There are classical indications for further psychological intervention when anxiety, depression, or psychosis immobilize daily activities or when a patient is posing special management problems. A nurse or social worker can handle many of these situations. Psychiatric back-up can be helpful for diagnosis and possibly medication. At some medical centers, psychiatrists, psychologists, nurses, and social workers are organized into a psychosocial team which works together with the physician in an ongoing monitoring role to provide intensified intervention when severe symptoms develop.

As knowledge of psychosocial issues in cancer becomes more developed, the possibility of screening cancer patients by psychosocial evaluation becomes more likely. Physicians will then be able to determine which patients have a greater propensity to develop emotional problems. Psychosocial care in cancer is effective both as a preventive strategy and as a therapeutic one. Acceptance of the need for emotional as well as physical care in cancer is increasing in the public's mind, and patients are accepting the value of such emotional evaluative services as an adjunct to treatment.

REFERENCES

Dunphy, E. J., Annual Discourse—On Caring for the Patient with Cancer. *New England Journal of Medicine*, **295**(6):313–319, (1976).

Holland, J., Psychologic Aspects of Cancer. *Cancer Medicine*, 991–1021 (1973).

Krant, M., et al., The Role of a Hospital-Based Psychosocial Unit in Terminal Cancer Illness and Bereavement. *Journal of Chronic Disease*, **29**:115–127, (1976).

Kubler-Ross, E., *On Death and Dying*. Macmillan, New York, (1970).

Mastrovito, Rene C., Cancer: Awareness and Denial. *Clinical Bulletin*, **4**(4):142–146, (1974).

Rosenbaum, E. H., *Living With Cancer: A Guide for the Patient, the Family and Friends*. Praeger Publishers, Inc., New York, (1975).

Schmale, A. H., Psychological Reactions to Recurrences, Metastases or Disseminated Cancer. *International Journal of Radiation, Biology, Physiology*, **1**:515–520, (1976).

Weisman, A. D., and Worden, J. W., Coping and Vulnerability in Cancer Patients. *National Cancer Institute Research Report*. (1977).

Weisman, A. D., and Worden, J. W., "The Existential Plight in Cancer: The Significance of the First 100 Days." *International Journal of Psychiatry in Medicine*, **7**:1–15, (1977).

Section V

Basic Concepts in Treatment Delivery

14

Techniques and Complications of Intravenous Therapy

1.0

Introduction

The intravenous route of drug administration is employed for many cancer agents for two reasons: 1) the physical characteristics of the drug require a specific formulation or 2) the drug is particularly unstable and may coagulate proteins which if administered intramuscularly will irritate the local tissues. Drugs which are unstable in acid solution cannot be administered orally because of the presence of acid in the stomach. Drugs which may be degraded, rapidly inactivated, or which cannot be transported across the gastrointestinal lining similarly must be administered by the intravenous route. Continuous infusion to promote adequate blood levels over protracted periods may be necessary if the drug is inactivated by the proteins or enzymes in the blood. The distinctions between infusion, perfusion, and bolus injection are reviewed in Chapter 4. For some drugs, only intravenous preparations are available.

The techniques for obtaining access to venous channels are crucial to the successful administration of drug therapy. This chapter reviews the techniques of routine intravenous access, discusses practical pointers for preserving long term venous accessibility, and presents new techniques or options for patients without venous access. The management of patients developing extravasation or infiltration is discussed, particularly with regard to surgical therapy of drug-induced cutaneous ulceration.

2.0

The Standard Intravenous Therapy Technique

The sequential steps in intravenous therapy are reviewed in the pictorial sequence (Figures 14.1a,b,c,d, & e). The first step is to identify the appropriate vein. This act is facilitated by obstructing flow proximally and allowing the vein to fill. The skin over the vein is prepared by alcohol and, in some instances, by an iodine solution, particularly in those situations in which blood cultures are to be obtained and superficial skin contaminants are to be avoided. Skin sterilization should be

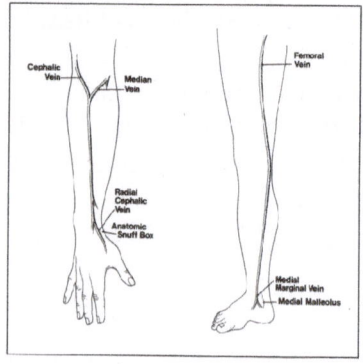

a

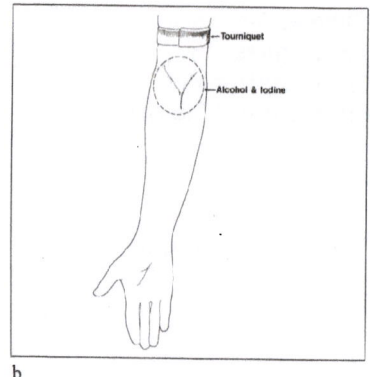

b

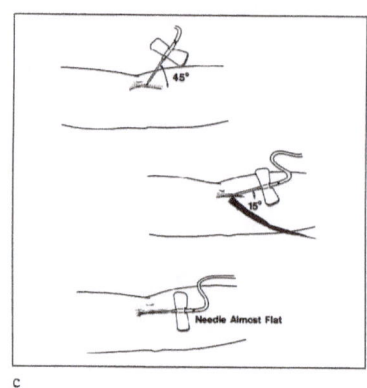

c

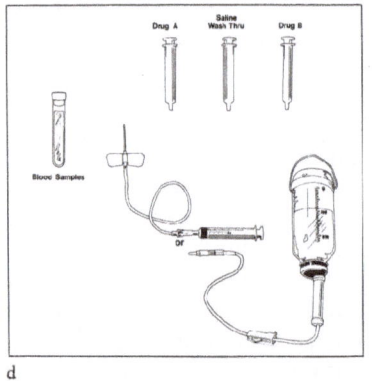

d

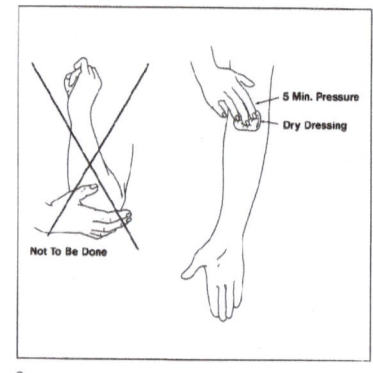

e

Figure 14.1
The 5 sequential steps in applying and introducing a needle are: (a) vein identification, (b) tourniquet application and sterilization of the skin, (c) needle insertion, (d) blood removal and drug administration, and (e) needle removal.

meticulous and proceed accordingly: alcohol followed by iodine, followed by alcohol again. The local area should then be allowed to air dry prior to needle insertion.

The crucial step is needle insertion. Prior to undertaking practice or trial runs in patients, intravenous therapy technicians should practice needle insertion in various fruits and other substitute substances. For the patient receiving intravenous chemotherapy, the use of scalp veins or butterfly needles is standard practice. In this manner intravenous therapy can be changed sequentially at a single sitting and patients may receive multiple drugs including transfusions through a single infusion set. In addition to butterfly needles, drip infusions may be employed if necessary for the particular drug program with the patient permitted some extra degree of mobility while receiving therapy over a 30 to 60 minute period.

The butterfly needle is prepared in a sterile fashion and the skin is drawn taut over the vein to be appropriated. The needle is placed with the bevel up and at a 45° angle to the skin overlying the vein. The needle is inserted under the skin quickly, then gradually into the wall of the vein in an almost parallel attitude to it. Slow insertion through the vein wall discourages extravasation or leakage of blood around the needle, and ensures that the needle will not pass through the opposite vein wall. With the needle in place, the butterfly is secured by applying non-allergenic tape over the needle site proximal to the puncture site avoiding hair contact. Additional tape is applied to secure the distal intravenous tubing as well.

204

Table 14.1
Chemotherapeutic Drugs Characterized by Sclerosing Effect on Veins

Vincristine

Actinomycin D

Mitomycin C

Adriamycin

Nitrogen Mustard

3.0

Table 14.2
Approach to the "Veinless" Patient (Non-Surgical)

1. Identify veins by palpation of standard sites (antecubital fossa)

2. Heating the extremity to dilate blood vasculature

3. Tourniquet manipulation

4. Local vein manipulation

The rate of injection of the drug depends upon specific drug characteristics including the volume of the drug to be administered (in turn dependent upon the solubility) and the local irritant effect of the drug. For drugs requiring a volume over 25 cc, it is recommended that they be dripped by infusion set. Drugs characterized by local irritation (see Table 14.1) should be injected slowly through an established intravenous flow of saline solution. Care should be taken not to allow the drug to accumulate in the intravenous tubing. Another important principle in the injection of intravenous drugs is the use of peripheral veins. For example, the peripheral veins on the dorsum of the hands can be used for initial therapy and then one can move proximally as these veins become less available. This technique ensures that there will always be proximal access. The exception to this basic principle is with the use of sclerosing drugs which should be injected primarily, if not exclusively, into large bore veins with high blood flows to dilute the local effects of the drugs. The large veins are most often identified proximally, for example in the antecubital fossa.

Finally, patients who will undergo long term intravenous therapy should have blood tests, such as hematologic complete blood count or liver function studies, obtained from finger sticks using microtechniques for the assays in order to avoid multiple venipunctures and risk the potential loss of venous access by trauma-induced thrombosis.

Dealing with the "Veinless" Patient
With increasingly successful treatment of cancer, patients are receiving therapy over protracted periods of time and living longer. Therefore, the availability of veins must be preserved. Often, however, constant exposure to drugs and venipunctures results in the depletion of adequate venous access or even the routine availability of veins. Venous access is primarily difficult in obese patients and in patients with hypoplastic veins.

Techniques of eliciting the unyielding vein to accessibility, and the development of alternative methods of direct venous access, are critical to patient management. The non-surgical approaches to the veinless patients are reviewed in Table 14.2. When no visible means of access is evident, palpable veins in standard areas, such as in the antecubital fossa, the anatomic snuff box, and over the medial malleolus, can be used. A second method is to apply local heat, dry or wet, although generally the latter, to the local extremity. This method induces dilatation of the arterial supply and maximizes blood flow locally. One important precaution is to ensure that the heat applied is not excessive so that second degree burns can be avoided, particularly in patients who have fragile skin and low platelet counts. Given these conditions, excessively heated extremities may lead to hemorrhage.

The process of increasing the blood pool in the veins by milking the veins from proximal to distal and by applying adequate tourniquet obstruction to encourage expansion of the vein blood pool is another technique. Finally, the identified vein can be dilated by gently striking the surface of the vein, inducing a proximal spasm resulting in blood pooling. The method is actually a sharp tapping of the vein with the finger and should be applied cautiously to avoid trauma-induced thrombosis.

4.0 Supplemental Methods in Intravenous Therapy

In the absence of adequate venous access by routine methods, one can occasionally resort to the oral route without compromising therapy. For many drugs, however, such as the anthracycline antibiotics and the periwinkle alkaloids, oral forms are not available or are still in the experimental stage. Furthermore, one is often concerned that the gastrointestinal absorption of such drugs may be suboptimal or less than adequate in circumstances in which the patient has had previous gastrointestinal surgery or has a disease which may directly affect absorption of the drug.

The alternatives to standard intravenous injection are primarily surgical as we see in Table 14.3. Administration of a drug by clysis or by continuous infusion into the subcutaneous tissues is occasionally possible with drugs which do not have a local sclerosing effect and which do not require a high volume of administration. Cytosine arabinoside is such a drug.

The use of peripheral arteriovenous shunts has been employed occasionally. Shunts developed for renal dialysis patients require a silastic connection between the artery and vein with the connector maintained outside the body. Major complications of these types of shunts result from clotting and local infection. Nonetheless, such shunts have been employed for the administration of continuous infusion chemotherapy and intermittent intensive chemotherapy. A related method of developing an arterial-venous shunt is that of establishing a direct surgical connection between the artery and the vein: an A-V fistula. The basic principle in the development of A-V shunts is that the venous channels become "arterialized" with the increased flow within the venous channel, resulting in secondary hypertrophy of the vein wall and dilatation of the vein channel. Such fistulae should be inserted early on in the following types of patients: 1) those exposed to sclerosing drugs; 2) those who will have therapy over protracted periods of time; and 3) those who have limited vein access when first seen. Development of shunts or A-V fistulae after extensive treatment with sclerosing drugs is often ineffective because of the failure to create a fistula. This failure is related in part to an inability to establish a venous connection due to sclerosis and thrombosis of the vein.

The use of surgical transplant grafts is another alternative. Swine artery, which has been radiated, may function as the graft. It has been extensively employed in dialysis patients and is actually a heterograft transplant of the carotid artery of the swine. Patients with cancer may have a hypercoagulable situation and be predisposed to develop thrombosis with such a graft. Therefore, a coagulation evaluation is essential.

Table 14.3
Alternatives to "Standard Intravenous" Injection

1. Clysis administration

2. Arteriovenous shunts (Scribner or Dialysis type)

3. Surgical heterograft shunts

4. Surgical arteriovenous fistulas

5. Subclavian catheter

Subclavian catheter lines are often placed for hyperalimentation and are useful in cancer therapy particularly for continuous infusion-type therapy. The subclavian line is not practical for patients receiving intermittent treatment or daily injection therapy because of the high risk of local infection, catheter failure, and thrombosis. In addition, this technique is not adaptable to outpatient management.

5.0

Therapy of Infiltration

A number of chemotherapy drugs are local venous irritants because of physical characteristics of the drug. For example, excessive acidity or alkalinity (pH alterations) of the reconstituted drug is not uncommon. Injection of any substance into veins is normally attended with disruption of the endothelial lining which results in clotting and thrombosis. Such effects may limit long term venous usage but, more importantly, extravasation outside the vein may result in major necrosis of the subcutaneous tissues, including muscles and tendons.

In order to avoid extravasation, it is particularly important to be aware of the drugs which have a potential for causing major tissue necrosis because not all drugs are irritants. The method of administration of these drugs is meticulous (Table 14.4). Such drugs should be diluted to reduce the drug concentration. The increased volume necessitates a longer period of infusion, but has less concentrated material which will impact locally on the vein wall. Injecting the drug into the side arm of a running intravenous is additionally helpful prophylactically as is the use of veins with large bores which have a high flow rate of blood. Sclerosing drugs should never be dripped by bolus into arterial or venous lines because the constant exposure of the local vein or arterial wall to the irritant may result in major local damage and thrombosis.

In the event of extravasation or suspected extravasation, the injection should be discontinued immediately and the intravenous line withdrawn. The local area should be marked and injected with a solution of xylocaine anesthetic and corticosteroid via a small bore needle or the Medijet gun applier. In addition, the limb should be raised and ice applied in order to promote resorption and to restrict arterial supply which results in secondary inflammation and fibrosis. The appearance of pain at the site of injection or local inflammation is a common phenomenon with all kinds of chemotherapeutic drugs, and the infusion of drugs should be merely slowed. If pain cannot be eliminated by slowing the rate of injection then the infusion should be discontinued altogether.

The only possible treatments for patients who develop severe local extravasation reactions are surgical debridement with split thickness or patch graft. Three patients with representative examples of extravasation injuries are demonstrated in Figures 14.2 and 14.3. The initial erythema may be delayed from 1 to 2 days after which central depigmentation indicating necrosis will develop (Figure 14.2). The lesion may then progress to necrosis of the entire subcutaneous thickness

Table 14.4
Administration Technique for Drugs with Local Sclerosing Effects—Avoid Extravasation

1. Dilute drug concentration in order to increase volume.

2. Inject drug into side arm of running i.v. fluid (be sure the fluid is compatible with the drug).

3. Use veins with high flow and large size.

4. Do not "drip by bolus" sclerosing drugs.

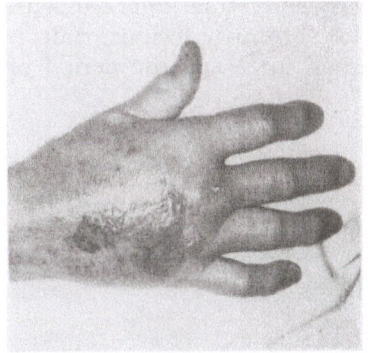

Figure 14.2
The drug administered on the dorsum of this hand resulted in localized pain without visible secondary effects until 7 days later. At this time, edema of the hand and erythema over the dorsum developed with ulceration and vesiculation developing at day 10.

207

with destruction of muscles and tendons and the development of secondary infection (Figure 14.3). Therapy for the intermediate stage is surgical debridement and superficial grafting (Figure 14.4).

6.0

Infusion Therapy

Some chemotherapeutic drugs must be administered over a protracted period of time so that they will be constantly available to the tumor cell. Continuous intravenous infusion will not only increase the drug's therapeutic activity but may also increase the patient's physical tolerance for the drug. For example, a dose of 5-Fluorouracil administered by bolus injection may be doubled or tripled when administered by continuous infusion before toxicity is observed. Some drugs, such as cytosinarabinoside, can only be administered by continuous infusion; others need to be combined with a large volume of fluid to make them sufficiently soluble for this process.

Continuous infusion is distinct from continuous perfusion. In the perfusion method the drug is administered through an arterial catheter, and the venous return from the area of arterial supply is occluded, thus ensuring that all of the drug delivered is sequestered and pooled in the area to which the artery is the primary or singular blood source. Arterial perfusion therapy is applied exclusively for extremity lesions but has occasionally been employed in the perfusion of head and neck tumors. In the latter instance, venous occlusion is obviously impractical. However, even in the extremity lesions, complete venous occlusion is rarely achieved and systemic dissemination of the drug is quite common. It has been proposed that if one could administer a high concentration of a drug directly by perfusion therapy without causing systemic dissemination, the effect on the bone marrow and other tissues would be minimized. However, because of the systemic leak of the drug into the venous system, myelosuppression and gastrointestinal effects occur. When arterial perfusion is used for extremity lesions, tourniquet occlusion is generally required for a specific period of time.

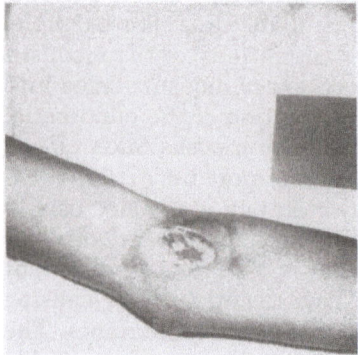

Figure 14.3
Surgical debridement is often necessary. In this example, 2 weeks after surgery, the lesion stabilized without continuing necrosis.

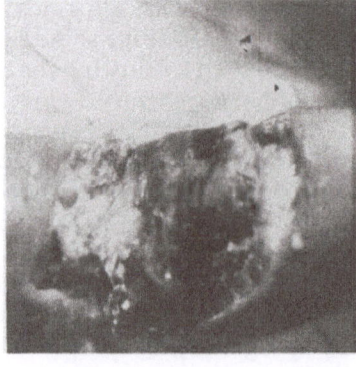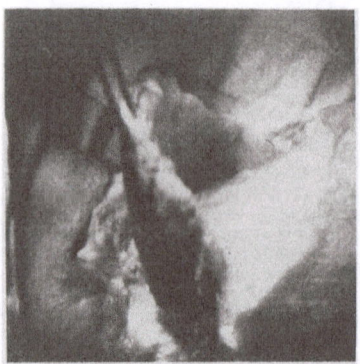

Figure 14.4
Depending upon the volume and quantity of drug administered, the necrosis may continue relentlessly to involve tendons, muscles, nerves, and the vascular supply. Secondary gangrene may develop, requiring extensive local debridement.

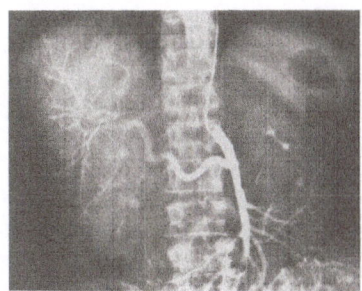

Figure 14.5
The catheter position in the hepatic artery is delicate and must be monitored by periodic arteriography.

Another form of arterial infusion is infusion of drugs into liver tumors. In contrast to peripheral limb perfusion, in which venous access is occluded, hepatic artery infusion has no venous occlusion and the infusion is continuous for days or weeks. Access to the hepatic artery may be achieved by abdominal exploration and surgical placement of the catheter into the hepatic artery. The catheter is subsequently attached to the external abdominal wall. Another method is the transcutaneous arterial approach which introduces the catheter through the brachial artery. The catheter is then positioned in the hepatic artery off the celiac plexus. An example of the catheter in the hepatic artery is illustrated in Figure 14.5.

For both the transcutaneous and the direct surgical approach, the catheter is connected externally to an infusion pump. The pump necessary for this procedure is either a chrono infuser (Figure 14.6) or a fixed floor model. The pocket chrono infuser gives the patient mobility, and the patient is generally able to adapt to the clock mechanism. Patients can be trained to make adjustments to the chrono infuser and to change medication bags when necessary, usually at five day intervals.

The drugs which have been administered by continuous hepatic infusion via an arterial catheter are generally 5-Fluorouracil and the deoxy derivative FUDR. Infusion methods employing adriamycin and other experimental drugs have also been applied. The rationale that infusion allows a higher concentration of a drug to be directed precisely at the tumor site is compelling, but definitive data as to the concentration of a drug extracted by the tumor is not available. The superiority of hepatic infusion therapy over standard venous injection therapy has not yet been proven. In centers in which the techniques of hepatic infusion are established, however, the tumor response rate to this method of drug delivery is substantial (50%).

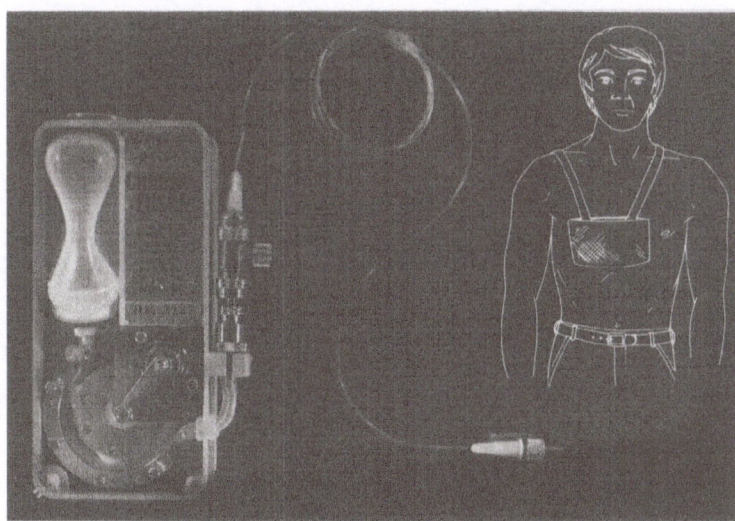

Figure 14.6
The chrono infusor is approximately the size of a clock and is worn in a convenient shoulder strap without major discomfort.

209

Table 14.5
Practical Pointers in Vein Maintenance

1. Use veins sparingly—obtain routine pretreatment blood tests via fingerstick method.

2. Use distal (hand) veins and in sequence advance up the arm; avoid veins at normal pressure sites.

3. Use large bore veins for sclerosing drugs.

4. Use butterfly needles for all infusions and particularly for multiple drug infusions.

5. Flush with saline following each drug injected to insure total dose is delivered and to avoid local mixed drug reactions.

6. Avoid local vein trauma; protect delicately placed intravenous lines with rings or other guard devices taped to the skin.

Infusion therapy has many complications. In extremity infusions, secondary venous occlusion may result in phlebitis; in arterial perfusions, arterial thrombosis and ischemic complications may develop. In the hepatic infusion program, arterial thrombosis is a common complication. Another problem is the stability of the catheter in the hepatic artery. If catheters are not maintained within the hepatic artery, high concentrations of the drug may be directed to tissues that are most sensitive to the drug, such as the gastrointestinal mucosa.

The role of infusion therapy in cancer management has not been established. A tumor which may be localized to a single visceral site can be effectively treated by local modalities, generally surgery and/or radiation therapy. Nonetheless, infusion therapy may be important for tumors which are specifically sensitive to chemotherapeutic management. It must still be established if tumors which are only marginally responsive to a drug may be more responsive when the drug is administered by direct infusion.

7.0

Summary

The meticulous care of venous channels is a critical part of cancer management today. Prophylaxis is the key factor, but establishing an awareness and a policy of vein management involving all members of the health care team including laboratory personnel is essential. The veins are literally the lines of life for these patients. Therefore, attention to the rules and guidelines in this chapter is an important aspect of cancer care. A summary of the practical pointers in vein maintenance is outlined in Table 14.5.

Index

*Italics indicate illustrations or tables.

213

Bursitis pills, *42*
Bursa-dependent cells (B cells), 91-92, *95, 96*
Bursa of Fabricius, 91
Busulfan, 48, *48,* 59, 180

Cachexia, 159, 165, 167-169, *169*
Cancer
  analogy technique and, 17, 19, 20
  bladder, *8, 9,* 76
  bone, *4,* 5
  brain, *64*
  breast, *4,* 5, 6, 7, 8, *49,* 50, *58,* 65-66, 77
  cervical, 3, 8, 28, 78-79
  choriocarcinoma, *9,* 75, 76
  colon, *4,* 5, 7, *8,* 10, 12, *92,* 131
  communicability of, 28
  endometrial, 78
  epidermoid, *49,* 78
  esophageal, 73-74, *75*
  gastrointestinal, 5, 8, 50, *56,* 81, 123, 178-179
  genito-urinary, *4,* 5
  head, *9,* 50, 74, *75*
  hereditary susceptibility to, 27-28
  larynx, 75
  liver, *64*
  lung, *4,* 5, 7, 8, 10, 12, 59, 78, 79-80
  lymphomas, 3-4, *4,* 5, 10, 51, 74, 80-81
  melanomas, *4,* 5, 7, 12, 51, 81
  myths about, 26-29
  neck, *9,* 50, 74
  oral, 73-74
  ovarian, *4,* 5, 6, 7, 8, 9, 48, *49,* 79, 131, 132
  pancreatic, *4,* 5, 147, 179
  pituitary, 79
  prostatic, *4,* 5, 8, 52, 53, 76
  rectal, 7, 8
  renal, *4,* 5, 10, 52
  reproductive system, 6, 78-79
  salivary glands, 74
  sarcomas, *4,* 5, 10, *49,* 77
  skin, 74, 81, 97-98
  staging for, *8, 8,* 76, 78-79, 80
  testicular, *4,* 5, 9, *49,* 75-76
Carbenicillin, 117
Carcinoembryonic antigen (CEA), 92-93
Cardiac effect, cytotoxic drugs and, 59
Career planning, 192
Catabolism, 166, 170
Cell growth cycle, 53-54, *53,* 55
Central nervous system, 77-78
Cephalosporin, 117
Cervical cancer, 3, 8, 28, 78-79
Cesium, radioactive, 70
Chemotherapy, 3-7, 8, 33-36, 39-41, 55-59, 56-57, *58,* 146
  anemia caused by, 114-116
  antiemetic therapy and, 126-130, *128, 129*
  blood brain barrier and, 10, 48
  combination, 60-61, 62-63
  gastrointestinal effects of, 124-125, *125*
  leukopenia and, 116-118
  marrow suppression by, 107, 109, 110-113, *111*
  nutrition and, 180
  tumor growth rate and, 10, 41, 53-55
  tumor response to, 4-5, *12,* 61-64, *62, 63, 64*
  ulceration and, 136-138, *139,* 140.
    See also Drugs, anti-tumor
Chills, 57
Chlorambucil, *48, 121*
Cholestyramine, 132
Choriocarcinoma, *9,* 75, 76
Chronic pain syndrome, 147-148
Chrono infusor, 209, *209*
Cirrhosis, 111
Citrovorum rescue method, 50
Clindamycin, 117
Clinical trials, 13-14, 39, 46-47, *47*
Closest person emotionally (CPE), role of, 189-190, 192, 193-194, 199
Clysis administration, 206, *206*
Cobalt, radioactive, 70
Colon cancer, 8, 10, 12, 92, 131
  chemotherapy and, *4,* 5, 7
Colostomy, 131, 147
Communicability of cancer, 28
Compazine, *129*
Conception, chemotherapy and, 36
Conconacalin A, 96
Consent form, 39, 40-41
Constipation, 34, *56,* 123, 130-131
Constitutional symptoms
  differential diagnosis of, 161-162, *161*
  treatment of, 162
  types of, 157-160, *158*
Contraception, 36, 59, 118
Contrast radiography, 72
Coolidge, W.D., 69
Cooperative group studies, 13
Cordotomy, 153, *153*
Corticosteroids, 52, *52,* 160, 162, 163
Cortisone, 198-199
Corynebacterium parvum, 98, 100
Cough medicines, 42
Coutard, H., 69
Craniopharyngiomas, 79
Cross-resistant drugs, 54-55
Curative therapy, 3, 4, 6, 73
Curie, Marie, 69
Curie, Pierre, 69
Cyclophosphamide, 48, *48,* 112, *121, 125*
Cylindromas, 74
Cyproterone, 52
Cystectomy, bladder, 76
Cystitis, 126
Cytosine arabinoside, *49,* 50-51, *121, 139,* 206, 208
Cytoxan, 7

Darvon, *42*
Daunomycin, 47, *47,* 59
Daunorubicin, *49*
Death, patient's interest in, 28-29, 196-197

214

216

Menopause, 36, 53, *57*, 58
Menstrual cycle, 36, 48, *57*
6-Mercaptopurine, *49*, 51, *121*
Methanesulfonates, *48*
Methanol extracted residue (MER) of
    BCG, 98, 99-100, *100*
Methotrexate, 7, 45, *45*, *49*, 112, *121*
  with citrorovum rescue, 50
  gastrointestinal effects of, *125*,
    138
    oral ulceration and, 136, 137-138,
      180
  pulmonary effects of, 59
  skin reactions to, 55
Methylcholanthrene (MCA), 93
Micrometastases, 7
Mind control
  for antiemesis therapy, 129-130,
    129
  for pain management, 153-154
Mithramycin, 49, *49*
Mitogens, 96, 97
Mitomycin C, 49, *49*, *205*
Mixed lymphocyte culture reaction
  (MLR), 97
Moderately responsive tumors, 4, 5
Monilia stomatitis, *138*, 139
Monocytes, 92, 96
Mononuclear cells, peripheral, 96-97
Mood elevators, *151*, 152, 158
MOPP therapy, 60, 80
Morphine, *151*, 152
Mucoepidermoid carcinoma, 74
Mucositis. *See* Ulceration, oral
Multiple drug therapy, 60-61
Multiple modality therapy, 6, 84
Mumps antigen, 95, 96
Muscle weakness, 35
Mustargen, 60
Mutagenesis, 48, 59
Mycostatin, 140
Myelosuppression, *48*, 51
Myths, cancer, 26-29

Narcotics
  for bowel function therapy, 130
  for pain, 150-152, *151*, *152*
  mentation effect, 199
Nail changes, *56*
Nasogastric feeding tube, 172
Nasopharynx cancer, 74
National Cancer Institute, 46
Natural products, therapeutic, 51,
  51, *121*
Nausea
  antiemetic therapy and, 126-130,
    128, *129*
  chemotherapy induced, 34, *56*,
    124-125
  metabolic imbalance and, 124
  nutrition and, 180
  radiotherapy and, 38, 126
Neck cancer, 9, 50, 74
Needle insertion, 204, *204*
Nembutal, *42*
Nephritis, 111
"Nerve pills," *42*
Neuraminidase, 101
Neurologic problems, chemotherapy
  and, 35, 51, *51*, *57*

Neutropenia, 116-117, 121
Neurosurgery, 152-153, *153*
Nitrogen balance, 166-167, *166*
Nitrogen mustard, 47, *48*, *121*, *125*,
  205
Nitrosoureas, 48, *48*, 112, *121*, *125*
Non-Hodgkin's lymphomas, 80-81
Normoblasts, 108
Null cells, 92
Nutrition
  assessment of patient and, 165-
    166
  cachexia and, 167-169
  chemotherapy and, 180
  oral pain and, 137
  protein metabolism and, 166-167,
    168-169
  radiotherapy and, 179-180
  support of patient and, 165-166
  therapy, *171*, 171-177

Oat cell carcinoma, 10, 78
Obstipation, 130
Obstruction bypass, 3, 4, 130
Oncogenic viruses, 92
Oncofetal antigens, 92-93
Oncovin, 60
Oophorectomy, 52
Oral cancer, 73-74
Oral ulceration
  chemotherapy and, 34-35, 136-
    138, *139*
  diagnosis of, 138-139, *138*
  predisposing factors for, 135-136
  treatment of, 140-141
Orchidectomy, 52, 76, 159
Oropharyngeal cancer, 73-74
Orthovoltage radiation, 70
Osteoarthropathy, pulmonary, 149-
  150, *150*
Ovarian cancer
  ascites in, 131, 132
  breast cancer and, 52-53
  chemotherapy for, 4, 5, 6, 7, 48
  radiotherapy for, 6, 7, *49*, 79
  staging system for, 8, 9
  surgery for, 79

Pain
  atypical, 148-150, *149*
  cancer types related to, 145-148
  clinical aspects of, 144-145
  hormone therapy for, 52
  management of, 3, 137, 150-154
  patient's fear of, 28, *29*
  perception of, 143-144
Palliation therapy, 3-4, 73
Pancreas dysfunction, 133, 147
Pancreatic cancer, 4, 5, 131, 179
Pancytopenia, 55, 107-109, 110
Paranasal sinuses cancer, 74-75
Paranoia, 27
Paregoric, 130
Parenteral nutrition, 175-177
Pathologic growth rate, tumor, 61
Patient
  isolation of, 20, 118, 186-187
  nutrition and, 165-177
  performance status of, 9, 158